TECHNICAL PAPER 17

ESTIMATES OF RESIDENTIAL BUILDING, UNITED STATES, 1840-1939

MANUEL GOTTLIEB

NATIONAL BUREAU OF ECONOMIC RESEARCH / 1964

TECHNICAL PAPER 17

ESTIMATES OF RESIDENTIAL BUILDING, UNITED STATES, 1840-1939

MANUEL GOTTLIEB
University of Wisconsin—Milwaukee

NATIONAL BUREAU OF ECONOMIC RESEARCH

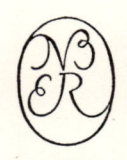

1964

Copyright © 1964 by
NATIONAL BUREAU OF ECONOMIC RESEARCH
261 Madison Avenue, New York, N. Y. 10016

All Rights Reserved

LIBRARY OF CONGRESS CATALOG CARD NUMBER: 64-15183

Price: $2.00

Printed in the United States of America

NATIONAL BUREAU OF ECONOMIC RESEARCH
1964

OFFICERS

Albert J. Hettinger, Jr., *Chairman*
Arthur F. Burns, *President*
Frank W. Fetter, *Vice-President*
Donald B. Woodward, *Treasurer*
Solomon Fabricant, *Director of Research*
Geoffrey H. Moore, *Associate Director of Research*
Hal B. Lary, *Associate Director of Research*
William J. Carson, *Executive Director*

DIRECTORS AT LARGE

Robert B. Anderson, *New York City*
Wallace J. Campbell, *Nationwide Insurance*
Erwin D. Canham, *Christian Science Monitor*
Solomon Fabricant, *New York University*
Marion B. Folsom, *Eastman Kodak Company*
Crawford H. Greenewalt, *E. I. du Pont de Nemours & Company*
Gabriel Hauge, *Manufacturers Hanover Trust Company*
A. J. Hayes, *International Association of Machinists*
Albert J. Hettinger, Jr. *Lazard Frères and Company*
Nicholas Kelley, *Kelley Drye Newhall Maginnes & Warren*
H. W. Laidler, *League for Industrial Democracy*
George B. Roberts, *Larchmont, New York*
Harry Scherman, *Book-of-the-Month Club*
Boris Shishkin, *American Federation of Labor and Congress of Industrial Organizations*
George Soule, *South Kent, Connecticut*
Joseph H. Willits, *Langhorne, Pennsylvania*
Donald B. Woodward, *A. W. Jones and Company*

DIRECTORS BY UNIVERSITY APPOINTMENT

V. W. Bladen, *Toronto*
Francis M. Boddy, *Minnesota*
Arthur F. Burns, *Columbia*
Lester V. Chandler, *Princeton*
Melvin G. de Chazeau, *Cornell*
Frank W. Fetter, *Northwestern*
R. A. Gordon, *California*
Harold M. Groves, *Wisconsin*
Gottfried Haberler, *Harvard*
Maurice W. Lee, *North Carolina*
Lloyd G. Reynolds, *Yale*
Paul A. Samuelson, *Massachusetts Institute of Technology*
Theodore W. Schultz, *Chicago*
Willis J. Winn, *Pennsylvania*

DIRECTORS BY APPOINTMENT OF OTHER ORGANIZATIONS

Percival F. Brundage, *American Institute of Certified Public Accountants*
Nathaniel Goldfinger, *American Federation of Labor and Congress of Industrial Organizations*
Harold G. Halcrow, *American Farm Economic Association*
Murray Shields, *American Management Association*
Willard L. Thorp, *American Economic Association*
W. Allen Wallis, *American Statistical Association*
Harold F. Williamson, *Economic History Association*
Theodore O. Yntema, *Committee for Economic Development*

DIRECTORS EMERITI

Shepard Morgan, *Norfolk, Connecticut*
N. I. Stone, *New York City*
Jacob Viner, *Princeton, New Jersey*

RESEARCH STAFF

Moses Abramovitz
Gary S. Becker
William H. Brown, Jr.
Gerhard Bry
Arthur F. Burns
Phillip Cagan
Joseph W. Conard
Frank G. Dickinson
James S. Earley
Richard A. Easterlin
Solomon Fabricant
Albert Fishlow
Milton Friedman

Victor R. Fuchs
H. G. Georgiadis
Raymond W. Goldsmith
Challis A. Hall, Jr.
Millard Hastay
Daniel M. Holland
Thor Hultgren
F. Thomas Juster
C. Harry Kahn
Irving B. Kravis
Hal B. Lary
Robert E. Lipsey
Ruth P. Mack

Jacob Mincer
Ilse Mintz
Geoffrey H. Moore
Roger F. Murray
Ralph L. Nelson
G. Warren Nutter
Richard T. Selden
Lawrence H. Seltzer
Robert P. Shay
George J. Stigler
Norman B. Ture
Herbert B. Woolley
Victor Zarnowitz

RELATION OF THE DIRECTORS TO THE WORK AND PUBLICATIONS OF THE NATIONAL BUREAU OF ECONOMIC RESEARCH

1. The object of the National Bureau of Economic Research is to ascertain and to present to the public important economic facts and their interpretation in a scientific and impartial manner. The Board of Directors is charged with the responsibility of ensuring that the work of the National Bureau is carried on in strict conformity with this object.

2. To this end the Board of Directors shall appoint one or more Directors of Research.

3. The Director or Directors of Research shall submit to the members of the Board, or to its Executive Committee, for their formal adoption, all specific proposals concerning researches to be instituted.

4. No report shall be published until the Director or Directors of Research shall have submitted to the Board a summary drawing attention to the character of the data and their utilization in the report, the nature and treatment of the problems involved, the main conclusions, and such other information as in their opinion would serve to determine the suitability of the report for publication in accordance with the principles of the National Bureau.

5. A copy of any manuscript proposed for publication shall also be submitted to each member of the Board. For each manuscript to be so submitted a special committee shall be appointed by the President, or at his designation by the Executive Director, consisting of three Directors selected as nearly as may be one from each general division of the Board. The names of the special manuscript committee shall be stated to each Director when the summary and report described in paragraph (4) are sent to him. It shall be the duty of each member of the committee to read the manuscript. If each member of the special committee signifies his approval within thirty days, the manuscript may be published. If each member of the special committee has not signified his approval within thirty days of the transmittal of the report and manuscript, the Director of Research shall then notify each member of the Board, requesting approval or disapproval of publication, and thirty additional days shall be granted for this purpose. The manuscript shall then not be published unless at least a majority of the entire Board and a two-thirds majority of those members of the Board who shall have voted on the proposal within the time fixed for the receipt of votes on the publication proposed shall have approved.

6. No manuscript may be published, though approved by each member of the special committee, until forty-five days have elapsed from the transmittal of the summary and report. The interval is allowed for the receipt of any memorandum of dissent or reservation, together with a brief statement of his reasons, that any member may wish to express; and such memorandum of dissent or reservation shall be published with the manuscript if he so desires. Publication does not, however, imply that each member of the Board has read the manuscript, or that either members of the Board in general, or of the special committee, have passed upon its validity in every detail.

7. A copy of this resolution shall, unless otherwise determined by the Board, be printed in each copy of every National Bureau book.

(Resolution adopted October 25, 1926, as revised February 6, 1933, and February 24, 1941)

Contents

	PAGE
Preface	xi
Foreword, by Moses Abramovitz	xiii
1. Introduction	1
2. Estimation of Decade Totals, 1890-1940	7
3. Estimation of Decade Totals, 1860-90	18
4. Estimation of Decade Totals, 1840-60	50
5. Estimation of Decade Annual Indexes	59
6. Evaluation of Total Series	82
Appendix	92

Tables

TABLE		PAGE
1.	Decade Estimates, Residential Units Built, United States, 1860-1940	8
2.	1940 Census Vintage Report on Dwelling Units Built, 1890-1940	10
3.	Percentage Change over Preceding Decade in Residential Building Aggregates, 1890-1920	16
4.	Selected Indexes of Comparability, Ohio and the United States, 1840-1910	19
5.	Percentage Distribution of Employment, by Major Industry Groups, Ohio and United States, 1950	23
6.	Structural Characteristics of Ohio and United States Dwellings, 1890-1940	25
7.	Demographic Comparsion of Ohio and United States, 1860-1910	27
8.	Net Household Formation by Nonfarm Labor-Force Increment and by Urban-Population Increment, Ohio and the United States, by Decade, 1860-1910	34
9.	Calculation of Estimates, National Nonfarm-Dwelling Increments, 1860-1910	36
10.	Labor Force, Farm and Nonfarm, United States and Ohio, by Decade, 1860-1910	38

Tables

		PAGE
11.	Derivation of Ohio Nonfarm Housing Estimates, by Decade, 1860-1910	40
12.	Derivation of United States Housing Estimates by Use of Ohio Multipliers, by Decade, 1860-1910	42
13.	Calculation of Adjusted Housing Production from Census Dwelling Increments, 1860-1910	44
14.	Array of Per Capita Residential Building Rates, Twenty-Nine States, 1840	55
15.	Estimated Production of Residential Housekeeping Units, 1840-1939	61
16.	Number of Urban and Other Dwelling Units Produced in Ohio, Annually, 1860-89	65
17.	Decade Indexes of Residential Building for the 1880's	68
18.	Decade Indexes of Residential Building for the 1870's	69
19.	Decade Indexes of Residential Building for the 1860's	70

Charts

	PAGE
1. Trend of Population in Ohio and the United States, 1900-50	20
2. Trend of Value Added by Manufacture, Ohio and the United States, 1899-1952	21
3. Trend of Electrical Power Production in Ohio and the United States, 1900-51	22
4. Covered Employment in Ohio and the United States, 1938-51	24
5. Average Size of Families in Ohio and the United States, 1860-1910	28
6. Ohio Male Population as Per Cent of United States Male Population, by Age, Census Years, 1860-1910	29
7. Ohio Female Population as Per Cent of United States Female Population, by Age, Census Years, 1860-1910	30
8. Decennial Per Cent Change in Urban Population, Ohio and the United States, 1840-1910	32
9. United States Decade Totals for New Dwellings, Three Variants, 1860-1910	43
10. Annual Mortgage and Building Activity, Nationwide 1880-89	48
11. Number of Persons Per House Erected in Ohio, by Counties, 1840, and Per Cent Growth of Assessed Real Property, 1841 over 1835	53

Charts

PAGE

12. Scatter Diagram of Percentage Growth of Population, Annually, 1835-40, and Per Capita Housing Production, Seven States, 1840   56

13. Decade Indexes of Dwelling Units, 1931-39   60

14. Decade Indexes for the 1890's   64

15. Decade Indexes of Urban Residential Building for the 1880's   71

16. Decade Indexes of Urban Residential Building for the 1870's   72

17. Decade Indexes of Urban Residential Building for the 1860's   73

18. Estimated Annual Residential Production, Constant and Shifting Weight Variants, 1860-89   74

19. Ohio Building, 1840-89   77

20. Decade Indexes for 1840-59   80

21. Estimates of United States Residential Building, 1840-1939   83

22. Number of New Dwellings, Ohio, 1857-1914   85

Preface

On the basis of fresh information and an extended use of Census sources, I have attempted in this research report to synthesize preceding statistical efforts to measure the annual flow of nonfarm residential building. I have, somewhat boldly, worked up the results in a new series of nonfarm residential production, running from 1840 to 1939. The series is anchored, at one end, in the 1840 Census count of dwellings erected and, at the other end, in the extensive probe into dwelling stocks disclosed by the 1940 Census. In the middle years, it is pinned to national projections founded on decade rates of residential building in Ohio, a new body of information presented in this paper, and also various urban building permit series already subjected to comprehensive analysis. The new estimates were thus derived by integrating results achieved by previous investigators with new information derived from a variety of sources.

The present version of the new nonfarm series is tentative only. At many points the analysis rests upon crude guesses with a range of possible error which could be narrowed by more intensive research. The range of possible error is widest in the representation of the course of short cyclical fluctuation. With respect to secular drift and long swings, I believe the new estimates are much more dependable.

The report is intended to be a contribution to our factual knowledge about United States economic growth and its tendency to long swings of about two decades in duration. It grows out of research during the past two years into all phases of urban building carried on with the aid of the National Bureau of Economic Research and particularly of Moses Abramovitz. We have in private correspondence discussed and wrestled with many of the issues developed in this report, which in turn grew out of a larger inquiry into the nature and characteristics of long swings in building activities. (See the progress reports on the inquiry in the Annual Reports of the National Bureau for 1962, pp. 48-51, and 1963, pp. 46-47.)

The results owe much to the devoted labor of my research assistants, Mary D'Amico and Asa Maeshiro. Mrs. D'Amico proved exceptionally

Preface

resourceful and diligent in organizing much of the detailed work summarized in this report. Thanks are also owed to friends who commented on an earlier draft and assisted the author in correcting some of its more obvious weaknesses. The charts were expertly drawn by H. Irving Forman, and the manuscript ably edited by Margaret T. Edgar. Geoffrey H. Moore made many helpful suggestions in his review. At an earlier stage, A. K. Cairncross of the University of Glasgow and David M. Blank, Columbia Broadcasting System, Inc., gave the manuscript the benefit of their counsel, and the review by Boris Shishkin, George Soule, and Willard L. Thorp of the NBER Board reading committee led to desirable textual revisions.

An exploratory study in New York City during the summer of 1960 owes much to the generous support of the Inter-University Committee for American Economic History. The research project was laid out during that study. Funds for the support of the study were provided primarily by the National Bureau of Economic Research. I am also grateful to the staff of the National Bureau for generous assistance in organizing the research, locating materials, and conducting tabulations. The processing and tabulation of the Ohio data were made possible by a special grant from the Rockefeller Foundation. A grant from the Wisconsin Urban Program (Ford Foundation) provided timely financial help in the summer of 1961 and made possible completion on schedule of our basic data collection. Finally, I am indebted to the officers of the University of Wisconsin for use of facilities and for grants of leave requested.

Foreword

The publication of Manuel Gottlieb's new study of United States residential building is one of those all-too-rare occasions when knowledge of an old subject of wide significance is advanced because a scholar has found and exploited new materials and has used older materials in new and useful ways. Previously, long-time series on the volume of building, some reaching back to the 1830's, have provided quantitative information bearing on several important phases of American economic history: the course of urbanization, the relation between population growth and economic activity, and the volume and composition of investment. They have also played an important part in establishing the fact that at least some aspects of American development were marked by long swings. First noticed in measures of building and real estate activity were the so-called building-cycles, with a duration of fifteen to twenty years. Such fluctuations have now been traced in many other spheres: immigration, railroad building, capital imports, incorporations, and the rate of growth of the money supply. Some scholars have advanced the view that they are to be found in the rate of growth of output at large. There is wide agreement that, in these large fluctuations of American development, a central role has been played by the long swings in construction and, within construction, by residential building in particular.

Study of all these questions unfortunately has been hampered by inadequacies in the statistical series on residential building and urban building generally. These failings are indicated in some detail in Gottlieb's introduction to the present study and elsewhere in the body of his report. The problem, in its essentials, derives from the fact that, during the entire period with which Gottlieb is concerned, residential building series rest chiefly on samples of cities in which permits to build were required. The data consist of the information derived from these permits on the number or value of residential permits or dwelling units or, in the earlier data, the value of buildings of any type. The cities included in the samples in the early decades contained only small portions of the

Foreword

total nonfarm population. By 1890, they covered about 15 per cent of that population, and in succeeding decades the coverage expanded rapidly. For the period since 1889, the series on which students now chiefly rely have been built up from the sample totals to estimates of all residential building. To do this, the samples were classified by city size or type, and estimating ratios derived from Census materials were used to raise the sample totals in each class to an estimated national total in each year. Before 1889, the restricted size of the samples did not permit such procedures. Instead, the series referring to earlier years[1] were constructed so as to provide only indexes of fluctuations and to eliminate, so far as possible, the effects of changing samples on the trend or annual movement of the series.

Close study has revealed shortcomings in the data for both the later period and the earlier. For the period since 1889, evidence has accumulated which indicates that existing series probably understate the volume of building, and unevenly in different decades. The possibility arises, therefore, that not only the volume of residential building but also its trend and major fluctuations are to some degree inaccurately depicted.

For the period before 1889, the existing indexes manifestly do not tell us what the volume of building activity was. They measure the growth of building only within the constant sample of cities which was at any particular time covered by the indexes.[2] They miss the growth that took place through the foundation of new communities and through the expansion of old ones beyond the legal limits of municipalities. They presumably misrepresent the fluctuations of building activity because the covered cities were, in general, the older and larger cities whose trends and fluctuations were not necessarily representative of younger and usually more rapidly growing communities.

Gottlieb's work makes a contribution toward overcoming all these difficulties. First, he has reconstructed the decade levels and, therefore, the trend of the data on the number of residential units built in the period since 1890. For this purpose, he has made use of the so-called vintage data provided by the Housing Census of 1940 together with other Census information about changes in the housing stock and population.

---

[1] Some of these series, such as those presented by Riggleman, Long, and Newman, also provided index numbers as far forward as the 1930's.

[2] The well-known Riggleman index does not even do that much, for it is expressed in terms of building per capita.

Foreword

Second, he has been able to place our knowledge of the level, trend, and fluctuations in residential building activity before 1890 on a new foundation, chiefly on the basis of hitherto-neglected records of construction and real estate activity in Ohio. These records, which Gottlieb may be said to have "discovered" and which he has worked up into usable form, provide a continuous and virtually complete annual record of new houses built in the entire state from 1857 to 1914. By making use of assessment data, Gottlieb was able to carry the Ohio series back to 1840.

The new Ohio series is in itself an important contribution to our knowledge of the growth and fluctuations of residential building and of the process of urbanization. For Ohio is an important state which, with its mixture of agricultural and growing industrial life, was to some degree representative of the country. In addition, however, Gottlieb has combined the Ohio information on building with data about the growth of urban population and the nonagricultural labor force to derive nationwide estimates of the level of building in successive decades. He has checked the estimates against Census information about increments to the urban housing stock adjusted for demolitions and losses, on the basis of information also obtained from the Ohio records. Thus we have for the first time a reasonable view of the level and trend of urban residential building stretching back to 1840. Next, Gottlieb has made use of the Ohio annual data to improve our picture of the year-to-year movements in nationwide residential building activity. As already indicated, the older series were based on samples of cities known to be less than adequately representative of all urban communities. The Ohio data, while limited to a single state, provide complete coverage of all residential building. Gottlieb has, therefore, combined the older series, which provide a more varied geographical coverage, with the new Ohio data, which are limited to a single state but cover residential building in communities of many sizes and types. He thus obtained annual indexes, which may be plausibly considered a better representation of year-to-year movements in building before 1890 than anything available up to now. Finally, Gottlieb applies his annual indexes to his decade averages to produce a continuous annual series adjusted to the level and trend shown by more comprehensive information.

Gottlieb's reconstruction of the data since 1890, the new annual series he provides from 1840 to 1890 and the Ohio residential building series, itself, provide us, therefore, with an essentially new picture of the level, trend, and fluctuations in urban residential building for the half-century, 1840-89, and with a picture changed in some important ways

Foreword

for the half-century, 1890-1939. As with all original work of this type, we cannot be sure at the moment of publication how well the new data will stand up. Close criticism of his sources and methods by experts are still needed to bring out the strengths and weaknesses of the figures. Needed too are repeated attempts to use the new material to illuminate the history of urban development and building fluctuations. How well it will be found to fit in with other related material is still to be seen. We can be confident, however, that Gottlieb's new data will play an important part in the continuing work of improving our quantitative knowledge of construction activity and will leave its imprint on a number of branches of economic history and analysis.

MOSES ABRAMOVITZ

1
Introduction

The history of efforts to give acceptable statistical representation to United States nonfarm building experience is full of endeavors which have illuminated or charted different aspects of a varied, a shifting, and an only partly explored field of building experience. Four different investigators have worked over building permit records of the nineteenth century and have prepared time series of urban residential or urban building: Riggleman (1934), Newman (1935), Long (1940), and Blank (1954). Others have prepared adjustments of the broadly based series of Riggleman and have developed variants: Isard (1947), and Colean and Newcomb (1952).[1]

All these efforts have been carried up to—in some cases beyond—the 1930's; five involved measurement of total urban building as recorded in building permits. Blank and Long provided separate measures of residential building permits, Blank in the form of number of dwelling units and Long in the form of number of residential permits. Both residential and nonresidential building have moved in long swings, but the rhythm and character of those swings differed somewhat in the two types of building, so that any series representing the value of total urban building must be used cautiously as an index of residential building.

[1] See John R. Riggleman, "Variations in Building Activity in United States Cities" (unpublished thesis, Johns Hopkins University, 1934); W. H. Newman, "The Building Industry and Business Cycles," *Studies in Business Administration,* University of Chicago Press, 1935, Vol. V; C. D. Long Jr., *Building Cycles and the Theory of Investment,* Princeton University Press, 1940; Walter Isard, "The Economic Dynamics of Transport Technology" (unpublished thesis, Harvard University, 1947); M. L. Colean and R. Newcomb, *Stabilizing Construction: The Record and Potential,* New York, McGraw-Hill, 1952, Appendix N, "Building Cycles," pp. 219-243; David Blank, *The Volume of Residential Construction, 1889-1950,* Technical Paper 9, New York, National Bureau of Economic Research, 1954.

Introduction

While these variant measures of our urban building history were being designed, overlapping series of measures were being worked out by other investigators for urban building activity since 1900, 1915, and 1920. Utilizing all available building permit materials and other information on construction, and using the technique of stratification of universe and checking against controlling census-derived figures, David Wickens developed a highly regarded and widely used residential building series for the period 1920-36.[2] Working independently, Chawner elaborated a residential building series going back to 1900 on the strength of permit records and Dodge contract information.[3] Official statistical agencies in the Departments of Labor and Commerce have developed residential and other building series carried back to 1915 and maintained currently. For the period 1889-1919, the official agencies have adopted the Blank residential building series, while, conversely, Blank and his National Bureau associates, Grebler and Winnick, have accepted the Wickens-Labor Department permit-derived residential unit series for the post-1919 period.[4] The 1889-1939 composite will be termed BLS-NBER series.

Meanwhile, Simon Kuznets and Robert Gallman developed over-all construction series going back quinquennially to 1840 and available on a smoothed annual basis since 1869. These measures are based upon decennial totals for construction materials produced in this country and destined for domestic use, with annual interpolators derived from annual series of particular types of building materials or quinquennial state census recordings of construction materials produced. The Kuznets and Gallman series show different rhythms and characteristics for some decades of the nineteenth century.[5]

The biases of all building-permit-derived series are now well known. Until recent years, those series, in effect, provided a record of building experience only within the covered central cities of metropolitan areas. Since building permits were not generally required in suburban and

---

[2] See final version as presented in David L. Wickens, *Residential Real Estate,* New York, NBER, 1941, pp. 41-50.

[3] L. J. Chawner, *The Residential Building Process,* Washington, 1939, and *Construction Activity in the United States 1915-1937,* Washington, 1938.

[4] *Nonfarm Housing Starts 1899-1958,* Department of Labor, Bull. 1260, 1959.

[5] See R. Gallman, "Commodity Output, 1839-1899," in *Trends in the American Economy in the Nineteenth Century,* Studies in Income and Wealth, Vol. 24, Princeton for NBER, 1960; Simon Kuznets, *Capital in the American Economy: Its Formation and Financing,* Princeton for NBER, 1961, Apps. B and C.

Introduction

satellite communities within metropolitan areas, the series did not catch noncentral-city building until those communities were annexed. Adequate allowance for the effect of annexations has always been difficult to make.[6] In addition, the patterns of building in central cities and in satellite and rural areas may diverge significantly. Adequate allowances for divergent rhythms of building in cities of different size classes have proved difficult to make. Finally, building has spread outward from city centers at a rate faster than the permit-reporting network has been broadened. These weaknesses have led to continual upward revisions of the more recent broadly based building-permit series. The permit-derived series for the nineteenth century, with their limited coverage, patently rest on insecure foundations.

The biases which run through construction-material-derived series are of a different character. First, amplitudes are dampened because of the substantial volume of construction materials used for maintenance and repair. Second, building and construction materials are used not only for building but also in other ways about which detailed information is known only for more recent years. Lumber is used for crates and boxes, as a fuel, in shipbuilding, and for manufacture of wooden products; bricks are used for sidewalks, street surfacing, and for underground construction.[7] Third, annual interpolators for the building-material series are comparatively scant and not representative during most of the nineteenth century. Finally, the most important building materials, lumber and brick, were produced through most of that century typically in small establish-

---

[6] For example, analysis of the results of the National Housing Inventory of 1956 disclosed that sizable undercoverage, which eluded the reporting permit network, was traceable to "failure to treat annexations correctly." At least during the period 1950-56, "surveys of non-permit housing starts seriously underestimated actual starts in non-permit areas." See *Progress Report on Improvements in Construction Statistics*, Census Bureau, Feb. 12, 1960, pp. 2 and 6.

[7] In 1912 some 74 per cent of lumber was used in construction while 26 per cent was used to produce box crates, furniture, vehicles, and other wooden products *(Lumber and Timber Products,* Works Progress Administration, GPO, May 1938, p. 108). In the nineteenth century wood was still widely used for heating and industrial fuel. Ohio railroads in 1858 used 209,416 cords of wood and 816,675 tons of coal *(Ohio Executive Documents,* Part 2, 1858, p. 584). In 1892 it was noted that "the utilization of brick for street paving has opened a new market for brick and created a distinct industry. Within the past few years thousands of miles of streets have been paved with this material throughout the West" *(Mineral Resources of the United States,* Bureau of Mines, GPO, 1892, p. 723); V. S. Clark, *History of Manufacturers in the United States,* McGraw-Hill for Carnegie Institution of Washington, 1929 ed., Vol. II, p. 494.

Introduction

ments, many of which were operated only part time to meet local needs.[8] The output of such establishments is in all countries difficult to evaluate statistically in reliable annual measurements.[9] The fourteenfold growth recorded by Gallman in the production of construction materials between 1840 and 1900 thus reflects in part the drying up of local production facilities, not fully reflected in our records.[10] Since the historic process of industrialization moved unevenly throughout the century, the patterns of movement of the census-recorded segment of the industry may give a deceptive account of total building activity.

At this juncture, new sources of information must be utilized. One helpful source may be found in the little-used "vintage" report of the *Housing Census of 1940*, which developed estimates of the decade in which 92 per cent of the enumerated dwelling units were erected.[11]

---

[8] "In the years preceding the Civil War the production process differed but little from the methods used since earliest historical times," with brick often being produced by hand methods and improvised kilns "at the building site" (A. J. Tassel, D. W. Bluestone, *Mechanization in the Brick Industry*, WPA, 1939, p. 4). The *1880 Census Weeks Report* commented (p. 27): "Brickmaking in many sections of the country is carried on upon a small scale and in a desultory way; the number of employees is small and the subdivisions of labor are of little importance." Even though William Haber reported that the brick industry between 1870 and 1890 had developed from "small scattered undertakings to a commercial enterprise of large proportions," still the Bureau of Mines in its 1895 report noted the difficulty of keeping a directory of producers up-to-date owing ". . . to the large number of plants, the constant establishment of new yards and the abandonment of old ones" (William Haber, *Industrial Relations in the Building Industry*, Cambridge, Mass., Harvard University Press, 1930, p. 25; *Mineral Resources*, 1895, p. 817).

[9] Right up to the present time," notes a Swedish account in the late 1920's, "the brick-making industry, included in the Industrial Statistics since 1873, only represents part of the total output of the country, as bricks are still largely manufactured as a home industry subsidiary to agriculture." Data with regard to the sawmill industry and the allocation of sawmill products to building and other uses was found too imperfect, before 1896, to warrant making annual estimates of value of construction using a "materials-used" base (E. Lindahl, E. Dahlgren, K. Kock, *National Income of Sweden 1861-1930*, London, P. S. King & Son, 1937, Part I, p. 177; Part II, p. 186).

[10] Gallman records a fourteenfold rise in construction materials used (millions of constant dollars) from 87 in 1839 to 1,224 in 1899. This contrasts with a sixfold rise in the recorded construction labor force from 269,000 gainful workers in 1839 to 1,640,000 in 1899. Gallman, "Commodity Output, 1839-1899," pp. 30, 63.

[11] See use of vintage returns by M. Reid, "Capital Formation in Residential Real Estate," *Journal of Political Economy*, Apr. 1958, pp. 135 ff.

Introduction

Another helpful source is the continuous census reporting of the decade increments of occupied dwelling units which, with appropriate adjustments for vacancy and shrinkage, should correspond in some fashion with estimated new residential production. The census decennial counts of incremental growth of urban population, occupied dwellings, and nonfarm labor force will, for any one decade, correspond to residential building only with a substantial margin of error, but the growth over a stretch of decades should aid in judging the adequacy of any set of decade housing estimates.

Reliable decade totals of new residential production cannot, however, be derived from the census alone. We need to use independent measures as well, in permit-reporting urban areas. Another measure has recently become available owing to the discovery of a new stretch of hitherto unutilized building and real estate data for an entire north central state, Ohio, available without lapses and with full coverage from 1857 to 1914. Annual increments (smoothed by a moving average) in the assessed value of town and city real property were found to correlate closely with annual statewide building values, permitting a projection of building back to 1840. We also have available for ten years in the middle seventies and early eighties an annual count, by number and value, of buildings in Ohio lost or demolished. This information sheds light on building shrinkage rates and thus helps to reconcile total production estimates with realized increments in standing structures.

Originally, the intention was to use the Ohio and other available materials to extend backward by a half-century the BLS-NBER series, which begins in 1889. Those estimates were regarded as sufficiently well established, in spite of the well-known general weaknesses of permit-derived statistics. However, our first efforts to utilize the new materials for extrapolation disclosed an appreciable gap between the aggregate of "starts" for the decades between 1890 and 1910 as measured in the BLS-NBER series and as projected by our materials. The series seemed to seriously overestimate production in the 1890's and to underestimate production in the 1900's. Yet, quite clearly, the estimates for the two decades could not be adjusted on the desired scale without upsetting the whole pattern of decade levels and the secular trend running through the pattern. Nor could we build up an annual time series for the decades preceding 1890 and pin it to a level that involved a serious overestimate. A continuous long series was required. Since other information also indicated that the BLS-NBER series tended to underestimation, the task of a general revision of our nationwide residential building statistics was under-

Introduction

taken, beginning with the task of extending statistical coverage backward in time.

The work of revision and extension is limited in this study to preparation of annual nationwide estimates of the number of new housekeeping permanent dwelling units erected. Our work of statistical revision and extension is also limited to preparation of estimates that will yield valid knowledge about long-swing movements. We did not find it necessary to achieve the high degree of accuracy in year-to-year measures that would be important for analysis of the short business cycle. Hence we have utilized methods of statistical adjustment designed to achieve tolerable accuracies in measures of growth trend and long swings, but which may not do justice to short cyclical movements.

2

Estimation of Decade Totals, 1890-1940

The first stage of the present research involves determination of a valid set of decade totals of nonfarm permanent residential housekeeping units constructed during the five decades following 1890. It begins with an appraisal of the relative adequacy of the decade estimates implicit in the BLS-NBER series, which begins with 1889. These decade estimates are the outgrowth of an extended series of studies in which data on building permits collected by permit-issuing municipalities were utilized. Two major undertakings sponsored by the National Bureau of Economic Research played an important role in the development of the series.[12] The decade estimates concerned are set forth in line 5 of Table 1. Elsewhere the table lists other sets of estimates related to housing construction. For the five decades following 1890, the total housing unit starts in the BLS-NBER series is 19,904,000 units.

For three reasons, this fifty-year total must be regarded as involving serious underestimation. First, Table 1 shows that the gross increases in nonfarm households and nonfarm stocks of dwelling units (lines 8, 9) recorded over the period 1890-1940 (19,952,000 and 20,179,000 units, respectively) are practically the same as in starts. This result is surely self-contradictory. It implies that permanent housekeeping units built were only as many as new households formed or the net dwelling stock increase between 1890-1940. No allowance is made for residential building which would permit more vacancies, would reduce household "sharing," or would replace loss in fires, disasters, or through voluntary demolitions.

---

[12] See Wickens, *Residential Real Estate;* L. Grebler, D. Blank and L. Winnick, *Capital Formation in Residential Real Estate: Trends and Prospects,* Princeton for NBER, 1956 (hereafter cited as GBW).

TABLE 1

Decade Estimates, Residential Units Built, United States, 1860-1940
(thousands)

Estimation of Decade Totals, 1890-1940

	1860's	1870's	1880's	1890's	1900's	1910's	1920's	1930's	1890-1940 Total
1. Blank, original estimates				2,940	3,610	3,590	7,497	2,650	20,287
2. Chawner, estimates					4,200	4,220	6,764		
3. Wickens, estimates				2,417	3,952	3,890	7,035	2,734	
4. BLS estimates, 1953						3,980	7,035	2,734	
5. Official BLS-NBER estimates				2,940	3,606	3,590	7,034	2,734	19,904
6. Survivors, 1940 vintage report			1,492	2,804	4,796	5,009	7,253	4,086	23,948
7. Official as per cent of vintage (line 5/line 6)				104.8	75.2	71.7	97.0	66.9	
8. Nonfarm household increments				2,351	3,858	3,469	5,700	4,574	19,952
9. Net increase, nonfarm dwelling stock				2,270	3,736	3,579	6,580	4,014	20,179
10. Census increments, occupied dwellings, adjusted for vacancy and shrinkage	974	1,112	2,482	2,021	3,409				
11. Wickens, estimated increment, private nonfarm families				2,262	3,445	4,109	5,541		
12. Ohio multiplier, nonfarm occupied dwelling increment	1,151	1,404	2,560	2,494	3,951				
13. Ohio multiplier, nonfarm labor force increment	938	1,221	2,553	2,411	4,318				
14. Decade estimates, new	1,061	1,333	2,597	2,491	4,200	4,220	7,497	4,019	22,427

Estimation of Decade Totals, 1890-1940

SOURCE, BY LINE

1. Blank, *The Volume of Residential Construction*, pp. 11, 59.
2. Chawner, *Residential Building*, p. 13.
3. Wickens, *Residential Real Estate*, p. 54.
4. *Construction During Five Decades,* Dept. of Labor, Bull. 1146, 1953, p. 3.
5. *Non-Farm Housing Starts 1889-1958*, Dept. of Labor, Bull. 1260, 1959, p. 15.
6. *Sixteenth Census of the United States, 1940, Housing,* Bureau of the Census, Vol. III, Part 1 (1943), Table A-1, p. 9 (except for exclusion of three months of 1940 as disclosed in the report in same series *Housing*, 1945, p. 3).
8. Grebler, Blank, and Winnick, *Capital Formation in Residential Real Estate*, p. 82.
9. *Ibid.*, p. 86.
10. See Table 13.
11. Wickens, *Residential Real Estate*, p. 55.

12 to 14. See Table 12 for 1860-1900 and Section 2, below, for later decades.

The starts total thus presupposes an implausible deterioration in housing standards over the fifty-year period.[13]

Second, critical scrutiny of later bench-mark studies indicates that, around the fourth decade of this century, our established permit-derived residential building statistics have tended to understatement ranging up to 20 per cent.[14] It is reasonable to hold that this tendency to understatement did not begin abruptly in 1940. Third, concrete indications of understatement are offered by the findings of the vintage statistics of the 1940 and 1950 Censuses. Owners, managers of residential buildings, or

---

[13] Though the tables were presented in GBW and commented on extensively, the issue of the consistency between cumulated starts and stock aggregates was bypassed. It was explicitly recognized that between 1890 and 1930, before the tendency to convert had allegedly grown strong, new calculated starts only slightly exceeded growth in households and dwelling stock (GBW, p. 88). The slight excess (of 988,000 units of housing stock) "reflects the net effects of demolitions, conversions and other changes in the housing stock" (p. 88). As Margaret Reid has pointed out, ". . . there will be a tendency to overestimate the number of conversions or to underestimate the number of demolitions in order to account for the net change in number of non-farm dwellings indicated by decennial censuses" (Reid, "Capital Formation," n. 23).

[14] Between 1930 and 1956, BLS estimates of housing starts "probably accounted . . . for between 70-80 per cent . . ." of the reported net change in units standing after liberal allowances for conversions and demolitions" (Grebler and S. Maisel, "Determinants of Residential Construction," mimeographed memo for Commission on Money and Credit, Oct. 1959, IV-17).

Estimation of Decade Totals, 1890-1940

knowledgeable neighbors were requested in the 1940 Census to disclose the year built of the original structure in which the surveyed residential units were located. The returns were presented by years for the first decade, by five-year intervals for the second decade, and thereafter by decade totals back to 1859. Vintage information was obtained for 92 per cent of the surveyed residential units.

Collecting decade totals for dwelling units for which year-built information was furnished, the vintage record (line 6, Table 1) as of the 1940 census enumerated 23,948,000 dwelling units in the 1940 stock as located in surviving structures originally erected after 1890. This total cannot be compared, without adjustment, with the 19,904,000 units estimated in the BLS-NBER series. Vintage attributions represent standing stock and thus include converted units, units transferred from the farm sector to the nonfarm and nonpermanent dwellings excluded from the starts category. The vintage attributions likewise do not include units built between 1890 and 1939 but destroyed or demolished, or for which a vintage report was not filed. Specified estimates under these headings are given in Table 2.

TABLE 2

1940 CENSUS VINTAGE REPORT ON
DWELLING UNITS BUILT, 1890-1940
(thousands)

Number nonfarm housing units originally constructed, 1890-1940 (unadjusted 1940 Census report)	23,948
1. Minus number of converted units included	2,436
2. Minus number of nonpermanent ineligible units included	147
3. Minus farm units transferred to nonfarm stock	150
4. Plus units built between 1890 and 1940 and destroyed or demolished	1,000
5. Plus units built after 1890 and not reporting vintage	1,805
Total adjusted	24,020

NOTE: For detailed explanation of the five adjustments, see the Appendix.

Estimation of Decade Totals, 1890-1940

The largest task of estimation involved the breakdown of converted units, numbering 3.18 million, and nonreporting units, numbering 2.35 million, into structures of origin erected before or after 1890. Some tendency was found for age to influence the distribution in both cases. Surviving older units are more prone to conversion, and it seems plausible that owners or managers of surviving older units are more likely to be ignorant of vintage. The evidence on hand indicated a stronger tendency for conversion to be correlated with age. We accepted the results of regression analysis, which put 76.7 per cent of the 1940 stock of converted units into the post-1890 vintage category. With less clear-cut evidence, the vintage category of nonreporting units was adjusted by the same percentage. With other adjustments, this produced an estimated total of 24,020,000 permanent housekeeping nonconverted dwelling units built between 1890 and 1940, or 20.7 per cent more than the number in BLS-NBER series. A detailed explanation of the adjustments set forth in Table 2 is given in the Appendix.

The table and Appendix presuppose that the vintage attributions are correct, or that errors of judgment or estimation as to year built were random. What are the probabilities of correct information and the direction of probable bias? In their rejoinder to Margaret Reid's use of vintage data along lines indicated above, Grebler, Blank, and Winnick pointed to the unreliability of vintage attributions for particular years.[15] For the individual years of the 1930's, the reports follow closely the index patterns derived from permit data (see Chart 13) except for a tendency to overestimation in 1930 and 1935. This is a manifestation of the well-known bias by which census age distributions cluster at multiples of 5. Census respondents should ordinarily have known the year-built of residential properties less than ten years old. Probably at least half the properties were resided in by people responsible for their original construction; only a small proportion would have passed though more than a second set of

---

[15] "Second, and more importantly, the number of dwelling units reported to the 1950 Census as being in structures built in, say, 1925 or even 1941 bears only a vague resemblance to the number of dwelling units actually built in 1925 or 1941" (Grebler, Blank, and Winnick, "Once More: Capital Formation in Residential Real Estate," *Journal of Political Economy*, 1959, p. 613).

owners.[16] Those owners, in turn, would have purchased with awareness of age.

If the decade of the 1930's passes muster, so can the decade of the 1920's. The total reported 7,253,000 units falls well within the range of the volume of starts estimated independently by Blank and by Wickens (see Table 1, lines 1 to 5). A respectable proportion of the properties were still lived in by people who would have had direct information about the timing of original construction.

As we turn to earlier decades, confidence in vintage attribution diminishes. The proportion of original or even second owners would have been much smaller. Age has a marked bearing on value, and abstracts of deeds available for scrutiny by owners or held by them usually indicate year of construction. But not all buyers commonly inspect abstracts, and census enumerators were instructed to accept estimates deemed reliable.[17] Under these circumstances, a bias toward over- or under-age estimation is possible. We can only check the returns for indications of any consistent bias cumulated in one direction.

---

[16] Tabulation of "year moved into" data from owner occupants in the 1960 Census shows that about half the owner occupants in Wisconsin have resided in the properties for ten years or longer. The average length of occupancy of an owner-occupied home is seven to ten years. (E. M. and R. M. Fisher, *Urban Real Estate*, New York, Henry Holt, 1954, p. 232). Nationwide census tabulation in 1960 of the urban population (including renters) showed that 23.9 per cent of the urban population had resided in the "present house" for ten years or longer or had always lived in the "same house." *Census of Population: 1960. General Social and Economic Characteristics, United States Summary*, Final Report PC(1)-IC GPO, 1962, Tables 71 and 72.

[17] The enumerators were instructed to find out the year built from an owner occupant, a well-informed neighbor, or a tenant. If the exact answer was not obtainable, the enumerator was instructed to enter "the approximate year based on available information and observation" (1940 Census, *Housing*, Vol. II, Part I, p. 195). The Census Bureau conducted no formal evaluation for this item. The responsible head of the Housing Division of the Bureau asserted: "From a qualitative standpoint we believe that this item [year-built data] is subject to rather large response errors, particularly for renter occupied units that have been built more than ten years prior to the date of the Census" (letter from D. B. Rathburn, Mar. 24, 1962). Systematic check of year-built census returns by census tracts in Milwaukee revealed that the returns tallied very closely with year-built returns of the independent real property inventory carried out between 1934 and 1936 (H. G. Berkman. *The Delineation and Structure of Rental Housing Areas*, University of Wisconsin Commission Reports, Vol. IV, 1956, p. 31, n. 5). So also a closely aligned vintage pattern was found between the 202 cities (1934-6) and 64 cities (1934) canvassed in the real property inventory surveys and the 1939 urban census enumeration (Peyton Stapp, *Urban Housing, A Summary of Real Property Inventories, 1934-1936*, WPA, GPO, 1938).

Estimation of Decade Totals, 1890-1940

Inspection of the decade patterns of vintage returns generally indicates a tendency to understate age for properties older than twenty years. Such understatement would cause the vintage attributions for the 1890's, 1900's, and 1910's to be improbably high in the light of other acceptable measurements. Thus, for the twenty years between 1900 and 1920, vintage records account for 9,805,000 units or 16 per cent more than the highest decade estimates from other sources (see Table 1). A tendency to understate age on older properties would result in vintage patterns showing high rates of implicit shrinkage or loss. Thus the 8,319,000 housing units of standard stock recorded in the 1890 Census became reduced to 3,220,000 vintage units recorded in the 1940 count. If nonreporting offsets conversion, the gap becomes the measure of loss through disaster or demolition. At that rate, housing stocks were halved in thirty-eight years or declined annually at a rate of 2.09 per cent, which seems higher than likely. At the same time there are distinct limits to the tendency to understate age. For older properties, features of style, location, method of construction, size of lot, and items of equipment give clues to the decade of origin. If we assume that the date was displaced a decade for 15 per cent of the vintage attributions of the 1910's and by rising percentage rates for later decades, properties labeled with post-1890 vintage rise by only 1.4 million units. Given the premise that the tendency to understating age commenced with properties older than twenty years of age and intensified with age, it becomes very difficult to account for more than 2 million units improperly shifted to the post-1890 category.[18] Hence we conclude that the vintage count of 1940 sets a limit, after adjustment for

[18] We used the following hypothetical schedule of underaging displacements, units in thousands:

Decade	Vintage Increment	Per Cent Displacement	Reduction	Increase	Corrected Decade Total
1910's	5,009	15	− 751	+ 751	4,258
1900's	4,796	25	−1,199	+1,199	4,348
1890's	2,804	50	−1,402	+1,402	2,601
1880's	1,492	75	−1,119	+1,402	1,775
1860-79	1,106	100%	−1,106	+1,119	1,119
Pre-1860	632			+1,106	1,738

Under this schedule about a third of the properties are displaced backward by a full decade; for properties older than forty years, the mean rate rises to near 70 percent. Yet only 1.4 million units are improperly shifted to the post-1890 age category.

Estimation of Decade Totals, 1890-1940

comparability, of underestimation in the BLS-NBER series of around 20 per cent, or some 4 million units for the fifty-year period, 1890-1940, or at the very least 10 per cent and 2 million units; and that any independently supported estimate falling within that range should be acceptable.

How should the BLS-NBER decade estimates for 1890-1940 be corrected for a tendency to underestimation ranging between 10 and 20 per cent? All decade totals could be scaled upward at a uniform rate. That would, however, presuppose the forces that biased the starts count to have worked uniformly over the decades concerned. A uniform bias is, however, unlikely in view of the varying coverage of permit-reporting areas and the unequal and shifting currents of rural, urban, and central city growth over the surveyed period. Hence, adjustment for underestimation has been based upon review, decade by decade, of the available evidence and judgment of the likely shifts in decade growth patterns.

For the decade of the 1930's we can accept with slight modification the verdict of the 1940 vintage report. For these relatively new properties, age estimates by census respondents should be reliable. Implicit annual rates of vintage production tallied closely with starts patterns (see Chart 13). Few of the newer vintage units of the thirties would have been converted, unreported for age, or wiped out by fire or demolition. Accordingly, we subtract from the vintage report only an appropriate allowance for nonpermanent or "ineligible" units, or for units built in the 1930's and transferred from the farm to the nonfarm sector.[19]

---

[19] In an unpublished note on "Naigles' Reconciliation of BLS Decade Starts with Census Stock Increments and Vintage Attributions," Moses Abramovitz allowed the following magnitudes (my estimates for the same items are in parenthesis).

Adjustments for Structures Built in the Thirties (thousands)
1. Conversions   70.8 (0)
2. Temporary and nonhousekeeping   141.6 (59)
3. Reclassification from farm sector   91.0 (8)
4. Demolition and other loss   − 4.0 (0)
5. Nonreporting vintage   −67.0 (0)

I follow Abramovitz in estimating that items 1, 4, and 5 substantially offset each other. My estimate for item 2 is the total number of ineligible units classified under the "other dwelling place" category and with a vintage traced back to the thirties. I have scaled down the possible reclassification from the farm sector to 8,000 units because the 1940 Census count disclosed that only 8.2 per cent of the rural-farm dwelling unit 1940 stock was built during the 1930's. If transference to the nonfarm sector was unaffected by age, then only some 8.2 per cent of the transferred rural-farm units were built in the thirties (see p. 95 below).

Estimation of Decade Totals, 1890-1940

For the decade of the 1920's we have available two independent efforts at measurement by Wickens and by Blank (7,035,000 and 7,497,000 units, Table 1, lines 1, 3). The sample of building-permit data utilized by both investigators was of the same magnitude, and both utilized refined estimating techniques. Blank commented quite properly that "external evidence affords no possibility of determining with any precision the degree of error in either of the two series."[20] However, since permit data tended to underestimation, it seemed reasonable to take the higher of the two estimates. Since the tendency to urban sprawl was inhibited in the twenties by the building splurge in central cities, permit statistics in the twenties were much closer to target than in the decade of the thirties with its marked drift of building outside the permit reporting-system.[21]

For the 1890's we have available decade estimates derived from Blank and from our own projections of the Ohio data (see Chapter 3). For checking, these results may be contrasted with census increments in household or dwelling stocks both unadjusted or as adjusted by Wickens or by myself (see Table 1). For various reasons the Ohio projections seemed preferable as an estimation basis for the 1890's. The Ohio-derived estimate tallies very closely with results reached by Wickens with the original census returns. Blank's sample of reporting systems started out in the 1890's with only 25 cities covering only 14.5 per cent of the nonfarm population; by 1900 sampled cities numbered 68 but with a population coverage of only 24.0 per cent.[22] The reporting sample was obviously too limited to permit refined estimation by urban size classes and regions. At the same time, the rapid growth of the sample may have generated bias. Finally, a basic assumption of the Blank expansion procedure is highly questionable, namely, that "nonfarm nonurban residential construction bears the same relationship to the increase in rural nonfarm population

---

[20] Blank, *Volume of Residential Construction,* p. 59.

[21] The acceleration of urban sprawl in the thirties is indicated by a variety of evidence. Thus population growth—and by inference residential building—was maintained during the thirties for the "rural ring" in metropolitan areas, though specific urban population growth fell off sharply (see GBW, p. 100). Likewise the 1940 Census vintage reports show that the urban segment of nonfarm building was steadily maintained within 3 percentage points of 80 per cent for the four decades after 1890 but fell to 63.4 per cent in the thirties. Since permit reporting systems provided weak coverage of building in the small towns or rural environs of central cities, the disparate behavior of permit-reporting systems in the twenties and thirties is explicable.

[22] Blank, *Volume of Residential Construction,* p. 35.

Estimation of Decade Totals, 1890-1940

that urban construction bears to the increase in urban population."[23] This overstates rural building by not allowing for the smaller urban-family size; it understates rural building on the other hand by not allowing for replacement building which would be unrelated to population growth. With these limitations, the Blank estimate for the 1890's seemed less acceptable than our Ohio-derived estimates. If the Blank level were accepted and given the decade patterns that seemed indicated, the aggregate level of building output for the eighty years after 1860 would be excessive in the light of end-1939 dwelling stocks and probable loss rates.

This leaves for judgment the two decades between 1900 and 1920, during which indications are that decade production was maintained substantially at the same level.[24] Likewise, indications are that the level in question involved an appreciable boost over the nineties and another boost to the twenties. Only two students, Chawner and Blank, have attempted a detailed canvass of the available records in order to derive an annual set of estimates on residential construction for the period. The Blank investigation was more thorough and involved a fuller set of building permit records; the Chawner estimates were prepared, however, under

TABLE 3

PERCENTAGE CHANGE OVER PRECEDING DECADE
IN RESIDENTIAL BUILDING AGGREGATE, 1890-1920

	1900's over 1890's	1920's over 1910's
New estimates (Gottlieb)	68.6	77.6
Blank	22.8	109.0
Wickens	63.5	80.8
Chawner		60.3
Census dwelling-stock increments	64.5	83.9
Ohio, statewide	67.3	

SOURCE: See Table 1, lines 1, 2, 3, 9, 14; and Table 11, col. 8.

[23] *Ibid.*, p. 48.

[24] See Table 1 lines 1, 2, and 3. Even the vintage magnitudes are within 5 per cent of each other.

Estimation of Decade Totals, 1890-1940

very competent direction. The Blank estimates level out at 3,600,000 units for the 1900's, the Chawner at 4,200,000 (see Table 1, lines 1, 2). As it happens, our Ohio-derived estimates for the decade of the 1900's run very close to Chawner's (4,135,000, Table 1, average of lines 12 and 13). The bias for underestimation of permit statistics would argue for the use of the higher of two independent sets of permit-derived estimates. Finally, such use in conjunction with the levels previously fixed for the nineties and the twenties yields a plausible set of decade-shift patterns (see Table 3).

For the five decades surveyed, our aggregate estimated production is 22,427,000 units or 12.8 per cent over the BLS-NBER aggregate. Our upward adjustments allow for about two-thirds of the gap between the vintage and the BLS-NBER aggregates.

3

Estimation of Decade Totals, 1860-90

We turn to estimation of decade production before 1890. For this task, the vintage report and urban-permit statistics give little guidance as to secular drift or decade shiftings. It is fortunate that new data have become available for nearly the entire second half of the nineteenth century.

The number and value of new buildings erected in the state of Ohio, by county, were reported annually from 1857 through 1914, together with marriage and real estate conveyance data. The nature of the statistical findings, the adjustments to which they were subjected, and the tests made of their validity will be reported on more fully in a monograph now in preparation.

Original collecting agents of the building schedule were local township assessors working under the direction of county auditors in a program of statistical reporting inaugurated by state law in Ohio in 1857. Local assessors, as a matter of official duty, would keep records of new building (and losses) for the purpose of maintaining property assessment rolls. We first ran audited tapes of reported county residential building and adjusted these tapes for deficient returns. The reported totals behave plausibly when contrasted with increments of assessed real property, or when laid out as time series. A variety of tests indicate that the data we use here have a high degree of reliability. Reports on nonpublic construction were in general compiled with more care than were reports on tax-exempt construction; and statistics, such as we deal with here, of dwelling units by number bypass the adjustments needed to allow for either the changing value of the dollar or shifting appraisal standards. The figures originally reported were adjusted only to allow for incomplete returns, for obvious errors in printing or arithmetic, and for conversion from a

Estimation of Decade Totals, 1860-90

record of "building completed" to a record of "building performance" for a uniform reporting period.[25]

Table 4 indicates that the state is qualified to serve as a basis for national estimation. The state was well settled by 1850 and responded fully to the building throbs of the middle-passage years of the nineteenth

TABLE 4

Selected Indexes of Comparability, Ohio and the United States, 1840-1910

	Percentage Growth Decennially in Population[a]				Urban Population as Percentage of Total Population		Percentage Share of Nonfarm in Total Labor Force		Total Nonfarm Income per Nonfarm Worker (dollars)	
	Ohio		U.S.							
	All	Urban	All	Urban	Ohio	U.S.	Ohio	U.S.	Ohio	U.S.
1840							22	21	356	437
1850	30	190	36	92	12	15				
1860	18	65	36	75	17	20				
1870	14	71	23	59	26	25				
1880	20	51	30	43	32	28	37	34	551	572
1890	15	47	26	57	41	35				
1900	13	32	21	36	48	40	56	44	609	622
1910	15	33	21	39	56	46				

Source: Richard A. Easterlin, "Interregional Differences in Per Capita Income, Population, and Total Income, 1840-1950," in *Trends in the American Economy in the Nineteenth Century*, App. A, pp. 97 ff.; *Ohio Population*, State of Ohio, Dept. of Industrial and Economic Development, 1960, Table 3; Bureau of Census, *1950 Census, Ohio*, pp. 35-36; *Statistical Abstract* 1920, p. 32; and *Historical Statistics of the United States, 1789-1945*, 1949, p. 25, Table B13-23.

[a] For decades closing in the specified year.

---

[25] Terminal dates of reporting years were either unspecified or shifted from July 1, May 1, and April 12. The practice was to make a spring survey (early or late) of what in effect was construction undertaken in the preceding year and completed by the reporting date. Some small structures could, however, have been commenced and completed within a reporting year ending May 1 or July 1. It was not possible to allow for this, and hence our calendar year allocations may have some "backward" bias. The adjustment for incomplete returns compensated for counties omitted from statewide returns. The adjustment was usually made by linear interpolation. Until the reporting system broke down in 1910-14, only a few counties were omitted from published returns in any given year.

Estimation of Decade Totals, 1860-90

century. Through that period, industry in Ohio was diversified and urban population was well distributed by size-class of city. In 1850 the state contained 8.54 per cent of the population of the country, 8 percent of manufacturing establishments, and 7 per cent of nationwide real estate value. Decennial rates of growth of urban population were falling both in Ohio and in the nation; and the share of total urban population was rising both in Ohio and in the nation. Table 4 shows that in terms of nonfarm income per nonfarm worker, Ohio by 1880 had moved close to the national average.

Craig and Yocum in a recent study note: "Over the past 50 years, Ohio's growth in population, industry, commercial development, transportation facilities, agricultural output—nearly any economic measure that can be taken—has kept pace almost precisely with the United States as a whole." Their three charts substantiating this finding are reproduced here as Charts 1, 2, and 3. They continue: "The reason there has

CHART 1
Trend of Population in Ohio and the United States, 1900-50

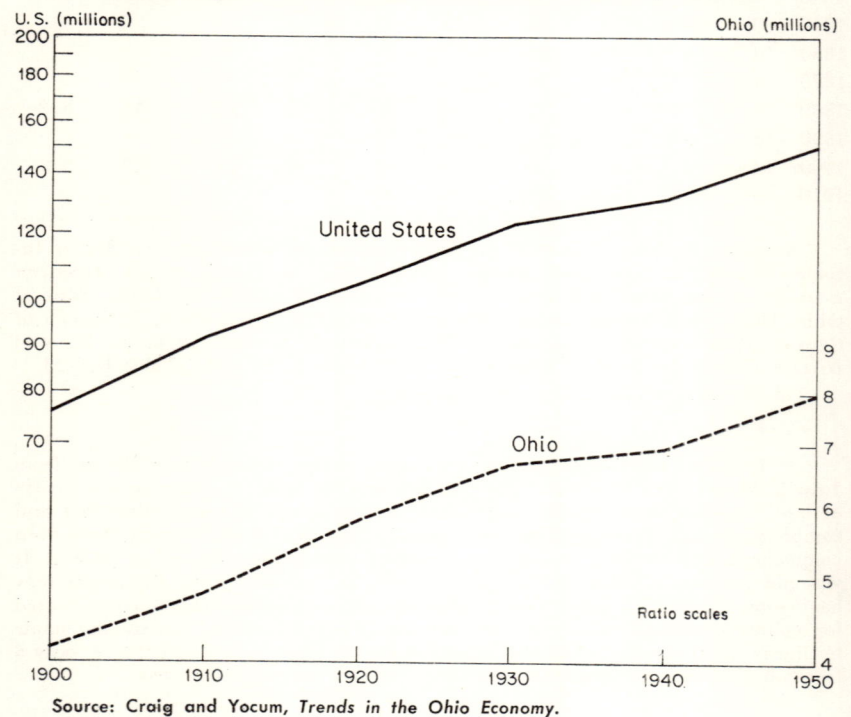

Source: Craig and Yocum, *Trends in the Ohio Economy.*

Estimation of Decade Totals, 1860-90

CHART 2

**Trend of Value Added by Manufacture,
Ohio and the United States,
1899-1952**

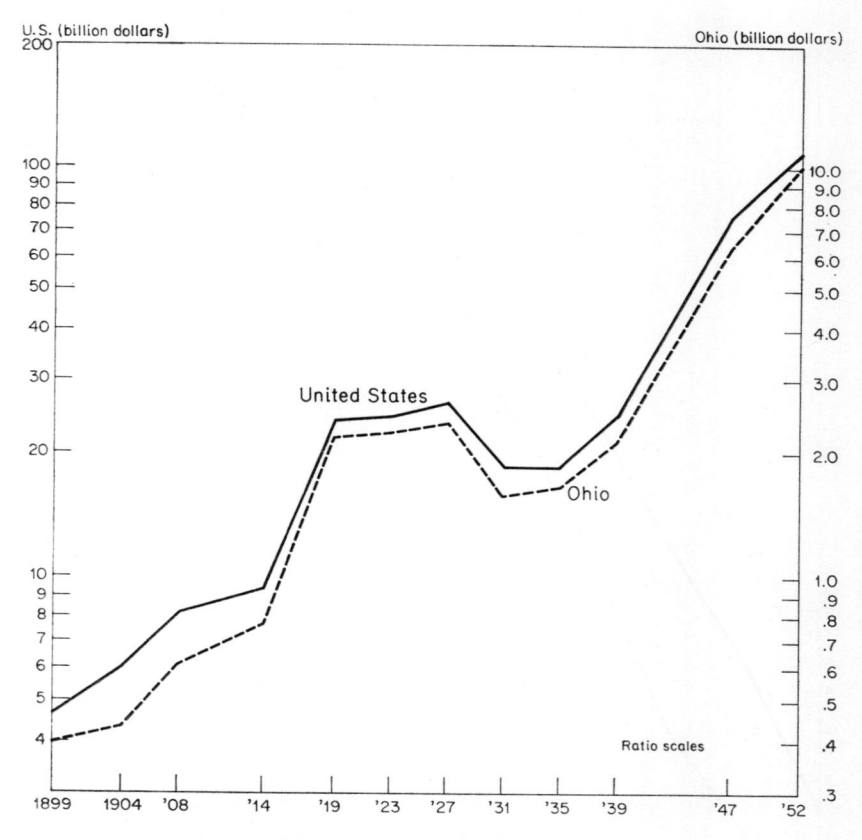

Source: Craig and Yocum, *Trends in the Ohio Economy.*

Estimation of Decade Totals, 1860-90

CHART 3

Trend of Electrical Power Production in Ohio and the United States, 1900-51

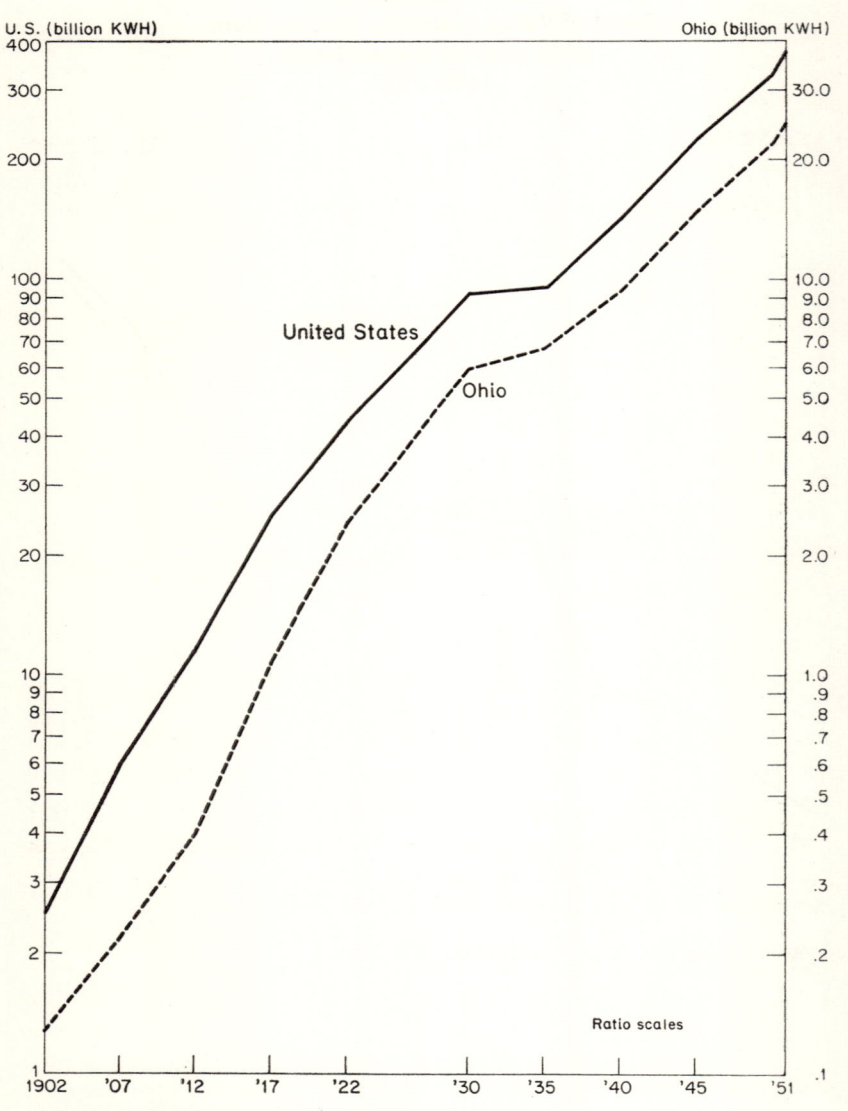

Source: Craig and Yocum, *Trends in the Ohio Economy.*

Estimation of Decade Totals, 1860-90

been a remarkable parallelism between Ohio and the United States as a whole lies not only in Ohio's central location, but in the combination of varied resources and circumstances which have permitted Ohio to develop a diversification of basic economic activities which closely mirrors that of the United States at large."[26] The study included an average index of deviation for 1950 showing divergence by selected states from the national distribution of employment by major industrial groups. The details for Ohio are shown in Table 5.

TABLE 5

PERCENTAGE DISTRIBUTION OF EMPLOYMENT, BY MAJOR INDUSTRY GROUPS, OHIO AND THE UNITED STATES, 1950

	Per Cent of Total U.S.	Ohio	Deviation of Ohio Percentage from U.S. Percentage
Agriculture, forestry, and fisheries	12.6	7.1	− 5.5
Mining	1.7	1.0	− 0.7
Manufacturing	26.3	37.1	+10.8
Transportation, communication, public utilities, and construction	14.1	13.2	− 0.9
Trade	19.0	18.5	− 0.5
Service and professional	18.3	16.2	− 2.1
Finance	3.5	2.8	− 0.7
Government	4.5	4.1	− 0.4
Total	100.0	100.0	21.6

SOURCE: Craig and Yocum, *Trends in the Ohio Economy,* Table 1, p. 12.

In terms of cyclical sensitivity, Ohio manufacturing is more concentrated than the national aggregate is in durable goods production (nationwide, 41.8 per cent of value added against Ohio, 59.3 per cent in 1939). Nevertheless, the estimated average per cent of compensable labor force unemployed in 1933 was nearly identical in Ohio (28.7 per cent) with the nationwide total (27.5 per cent). The state has also shown employment trends almost parallel to those of the United States (see Chart 4).

[26] P. G. Craig, J. C. Yocum, *Trends in the Ohio Economy,* Bureau of Business Research, Ohio State University, Res. Mon. 79, 1955, p. 1.

Estimation of Decade Totals, 1860-90

CHART 4

Covered Employment in Ohio and the United States, 1938-51

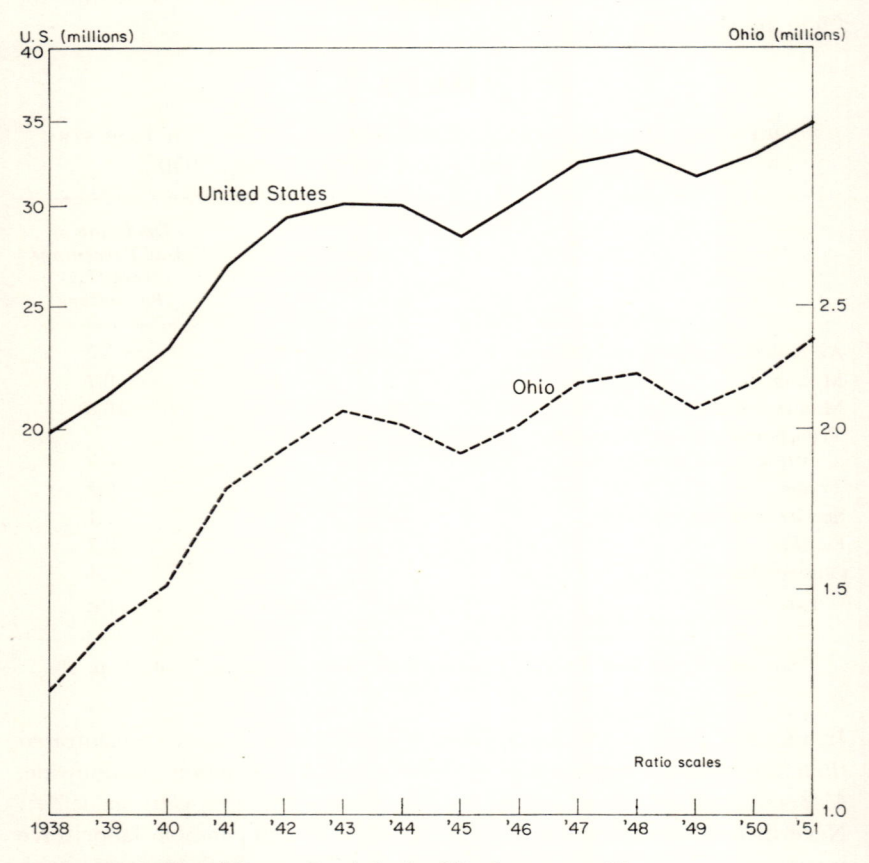

Source: Craig and Yocum, *Trends in the Ohio Economy*, p. 29.

Estimation of Decade Totals, 1860-90

Some structural characteristics of nationwide and Ohio dwellings, 1890-1940, are given in Table 6. By and large, the layout of characteristics is reassuring. The distribution by number of rooms is modal at the five- to six-room house, though there are fewer small units and more larger units in Ohio. The per cent of rented dwellings was somewhat less than

TABLE 6

STRUCTURAL CHARACTERISTICS OF OHIO AND UNITED STATES DWELLINGS, 1890-1940

Percentage Distribution 1939 Nonfarm Dwelling Stock by:	Ohio	U.S.	Percentage Distribution 1939 Nonfarm Dwelling Stock by:	Ohio	U.S.
Number of Rooms			City-size class (continued)		
1	2.3	3.7	Built 1900-40	54.8	73.4
2	5.4	8.4	Built before 1900	39.4	20.6
3	9.9	14.2	In converted units	7.5	7.6
4	13.8	18.1	Rural farm, entire	16.6	25.8
5	23.3	20.7	Built 1900-40	39.5	69.5
6	24.6	17.7	Built before 1900	55.7	26.5
7	9.8	7.4	In converted units	3.8	2.9
8 or more	9.9	8.4	In PMD, entire	62.6	55.5
Not reporting	1.0	1.3	Built 1900-40	72.8	72.3
Period built			Built 1879-1900	16.4	14.1
1935-39	4.7	7.9	Built before 1879	6.2	5.0
1930-34	4.0	6.0			
1925-29	12.4	13.5	Per cent of nonfarm units rented		
1920-24	10.5	11.1			
1910-19	18.0	17.0	1930	48.50	54.56
1900-09	17.3	16.2	1890	54.64	63.10
1890-99	11.5	9.5			
1880-89	7.0	5.1	Percentage 1940 nonfarm units, 1-family detached	62.0	55.2
1879 or earlier	8.9	5.9			
Not reporting	5.5	8.0	Average value, all nonfarm dwellings		
Internal features			1930	$5,138	$5,022
In converted units	10.2	9.3	1900	1,671	1,951
Converted to residential	1.1	1.4			
With private bath and flush toilet	60.3	57.6	Average value, mortgaged nonfarm dwellings		
With central heating	62.1	46.0	1890	$2,366	$3,250
City-size class			1920	5,012	4,938
Urban, Outside PMD[a]	21.4	22.7			
Built 1900-40	57.3	68.7	Average value, nonfarm home mortgage		
Built before 1900	34.3	22.9	1890	$ 879	$1,293
In converted units	12.1	13.3			
Rural nonfarm, Outside PMD[a]	16.0	21.9			

SOURCE: Wickens, *Residential Real Estate*, pp. 80-85, Tables A-1, A-3; *Sixteenth Census, 1940, Housing, Characteristics by Type of Structure*, Tables A-1 to A-5, pp. 3 ff., 270-289; Dept. of Interior, Census Division, *Report on Farms and Homes*, 1896; Census Bureau, *Wealth, Debt and Taxation*, GPO, 1907, p. 17.

[a] PMD = principal metropolitan district.

the U. S. figure and the percent of detached one-family dwellings somewhat greater. Possibly the larger size is offset by the older age as reflected in the age distribution. In terms of value, Ohio units were in 1930 only 2.3 per cent above the nationwide average.[27] Conversion rates in Ohio closely paralleled the rate in the nation. Except for a lesser farm and a greater principal metropolitan district (PMD) share, the size-class patterns are close. The share of nonmetropolitan urban is nearly identical for Ohio and the nation (21.4 per cent and 22.7 per cent respectively). The rate of growth of housing stock within the Ohio PMD's matches the national pattern. The rate of growth of housing stock in Ohio outside PMD was faster, however, than nationwide before 1900 and slower thereafter, particularly for rural nonfarm areas and to a lesser degree for urban areas outside PMD.

Table 7 and Chart 5 present data indicating comparability between Ohio and the nation with regard to demographic and housing characteristics for the years 1860-1910. The average Ohio family size fell below the national level from 1880 on, reflecting the higher degree of urbanization in Ohio and relatively greater numbers of the typically smaller nonfarm family. The smaller family size so far as children are concerned is indicated by the age distributions. The proportion of productive members of the population—the 15-60 age class—was virtually the same for both Ohio and the nation throughout the entire period. The smaller family size due to fewer children is offset by greater longevity. Rates of marriage for Ohio and the nation were nearly identical.

Charts 6 and 7 spell out the population profile of Ohio in census years, 1860-1910, expressed as percentages of the national totals by age brackets. The steady drop of profile curves for later years reflects the slower rate of population growth in Ohio and the post-1880 smaller Ohio family size. The profiles also show traces of a migratory wave between 1850 and 1880 and a resulting predominance of aged over young. That wave drained away young men—and to a lesser degree women—in the productive and fertile age brackets of 15-35 years. The tendency to emigrate out of the state was apparently sustained through the 1870's. By 1890, selec-

---

[27] This finding as of 1930 is only apparently deviant from the fact that, in 1890, Ohio mortgaged nonfarm homes were some 27.2 per cent short of the nationwide average. (See Table 6, lines 5 and 6). For, as we shall see later, Ohio residential building in the middle eighties was relatively depressed, while residential building elsewhere boomed. Thus, in 1890, the ages of mortgaged homes and mortgages in Ohio were relatively greater than nationwide ages were. We shall return to this issue in a later study dealing with value levels.

Estimation of Decade Totals, 1860-90

TABLE 7
DEMOGRAPHIC COMPARISON OF OHIO AND UNITED STATES, 1860-1910

CENSUS YEAR	Average Size Family Ohio	Average Size Family U.S.	Ohio as Per Cent of U.S.	0 to 14 Ohio	0 to 14 U.S.	Ohio as Per Cent of U.S.	15 to 60 Ohio	15 to 60 U.S.	Ohio as Per Cent of U.S.	60 and Over Ohio	60 and Over U.S.	Ohio as Per Cent of U.S.	Ohio, Estimated Net Intercensus Migration[a] (thousands)
1860	5.39	5.28	102.1	41.20	40.50	101.7	54.25	55.05	98.6	4.54	4.45	102.0	
1870	5.11	5.09	100.4	39.26	39.20	100.1	55.21	55.77	98.9	5.53	5.03	109.9	
1880	4.98	5.04	98.8										−12.9
1890	4.68	4.93	94.9	32.95	35.52	92.7	58.66	58.03	101.1	8.39	6.43	130.4	41.9
1900	4.40	4.76	92.4	30.85	34.47	89.4	61.21	59.10	103.5	7.94	6.43	123.4	77.7
1910	4.20	4.54	92.5	28.19	32.13	87.7	62.89	61.10	102.9	8.92	6.77	131.7	207.7

PERCENTAGE DISTRIBUTION AND MARITAL STATUS, WOMEN 15 YEARS AND OVER

CENSUS YEAR	Single Ohio	Single U.S.	Ohio as Per Cent of U.S.	Married Ohio	Married U.S.	Ohio as Per Cent of U.S.	Percentage Rate of Marriage, Females, 15 Years and Over Ohio	U.S.	Ohio as Per Cent of U.S.
1890	32.4	31.8	101.8	56.5	56.8	99.4	3.126	3.072	101.7
1900	31.1	31.2	99.6	57.2	57.0	100.3	3.006	3.147	95.5
1910	28.9	29.7	97.3	59.5	58.9	101.1			

Estimation of Decade Totals, 1860-90

CHART 5

Average Size of Families in Ohio and the United States, 1860-1910

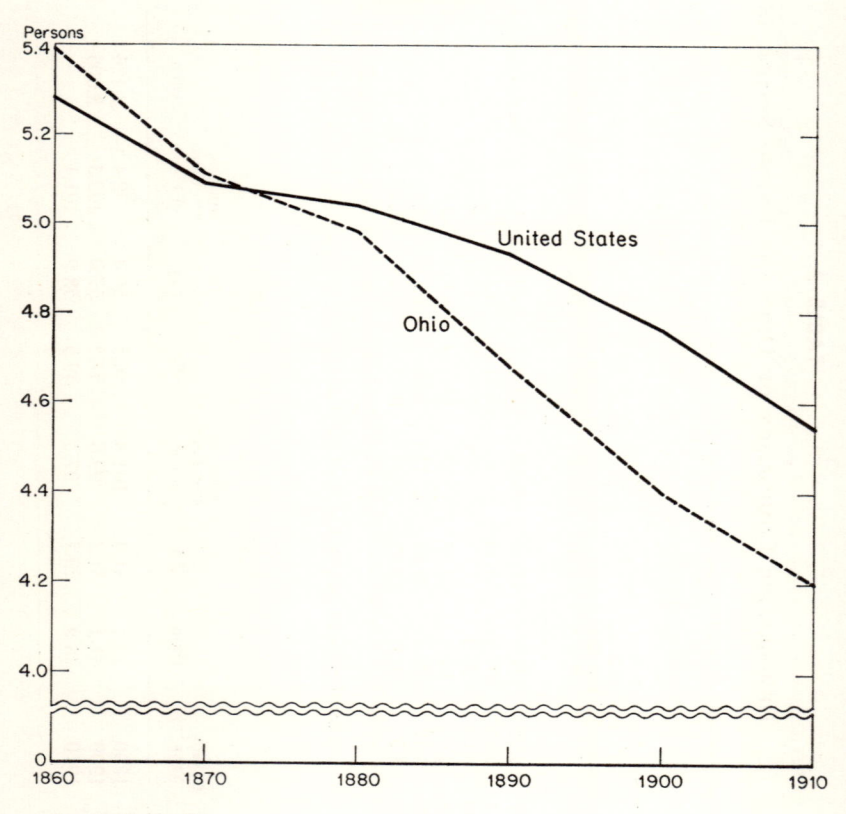

Source: Table 7.

Estimation of Decade Totals, 1860-90

CHART 6
Ohio Male Population as Per Cent of United States Male Population, by Age, Census Years, 1860-1910

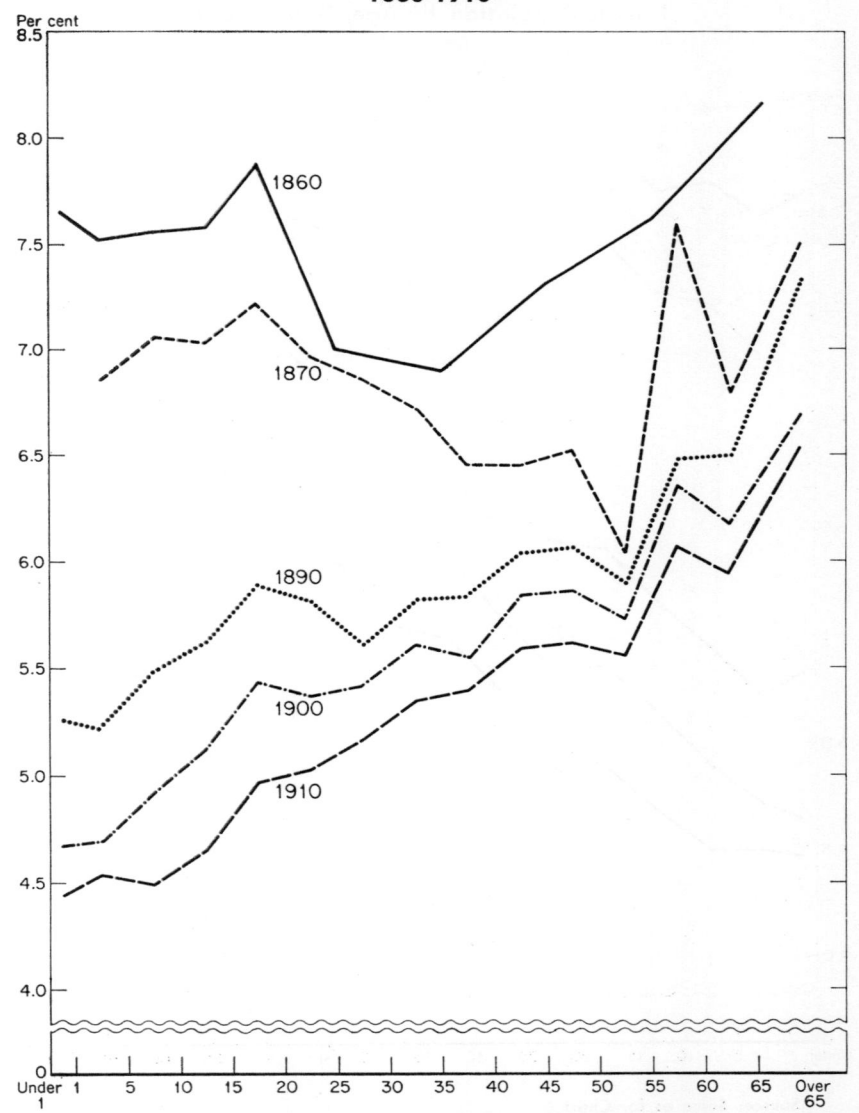

Source: 1860, 8th Census, *Population*, pp. 370-371, 596-597. 1870, 9th Census, *Vital Statistics*, pp. 580-593. 1890, 11th Census, *Population*, Part II, pp. 72-73, 2-5. 1900, 12th Census *Population*, Part II, pp. 78-79. 1910, 13th Census, Vol. 1, *Population*, pp. 303, 394.

Estimation of Decade Totals, 1860-90

CHART 7

Ohio Female Population as Per Cent of United States Female Population, by Age, Census Years, 1860-1910

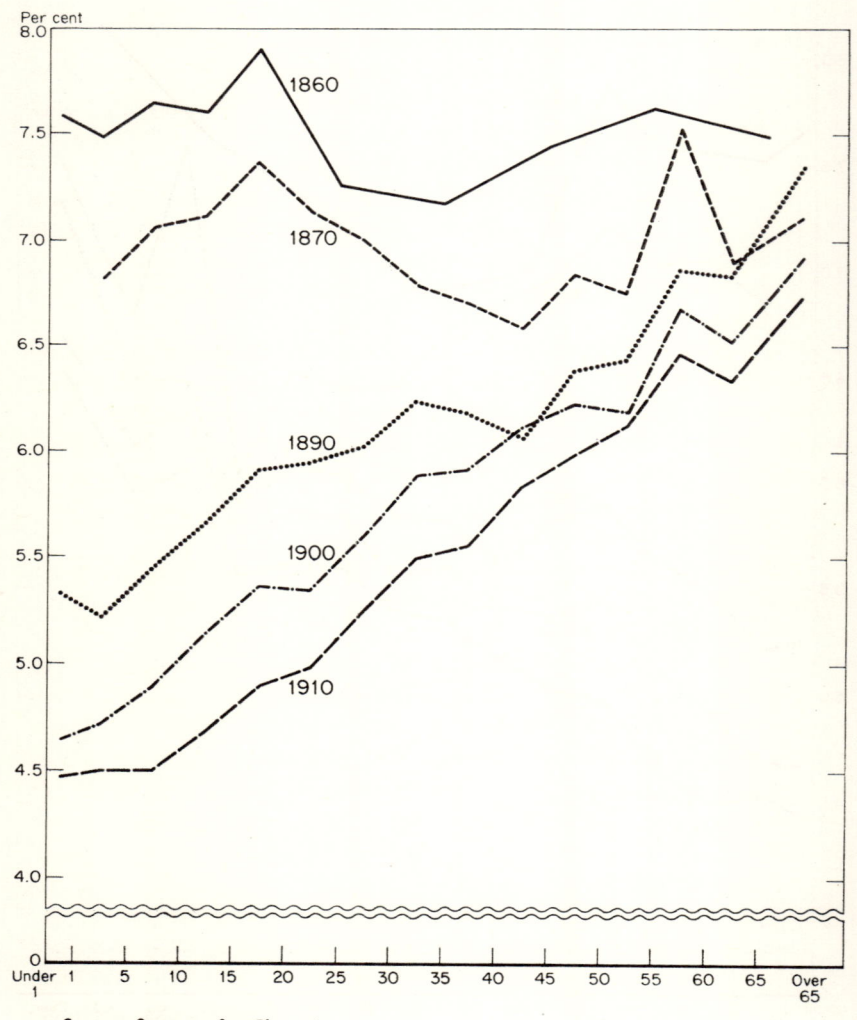

Source: Same as for Chart 6

Estimation of Decade Totals, 1860-90

tive age emigration had become mild, and by 1900 and 1910 it had disappeared. Net emigration was apparently reversed during the 1870's (see Table 7) with steadily rising net intercensus immigration. Immigrants apparently account for the Ohio tendency to a higher average age.

These materials indicate that while the state may readily serve as a base for national projection, allowance must be made for certain peculiar features of the state's economic development. Thus, in terms of percentage decennial growth of urban population, inspection of Chart 8 shows three nonconforming decade movements. The decline in the rate of urban growth between 1850 and 1860 was much steeper for Ohio than for the nation. During the sixties, Ohio urban population growth, unlike that of the nation, reversed trend. During the eighties, Ohio urban population had little of the booming growth that marked the national course, indicating that the comparative intensity of long urban-building swings experienced in Ohio cannot be mechanically projected to the national scene.

While Ohio's urban growth and industrial development broadly matched in intensity the nationwide movement, its agricultural population and settlement showed comparatively little growth after the Civil War. Hence, the state became more highly urbanized and industrialized than the nation as a whole. This, in turn, along with selective age migration, helped produce a relatively smaller average household size and a household age composition different from the nation's.

Three alternative sets of census increments are available to project Ohio building experience into nationwide aggregates: urban population, nonfarm dwelling stock, and nonfarm labor force. The first two are derived from the census enumeration of population and dwellings; the third, from the census enumeration of occupations. The three projection bases represent different facets of the process of economic and demographic growth. Growth of the employed nonfarm labor force would presuppose additional households and dwellings. If the rates of household formation per unit of nonfarm labor force are the same in Ohio and in the nation, and if additional nonfarm households are similarly apportioned between urban and nonfarm rural locations in Ohio and the nation, then all three projection bases should yield identical returns. But divergences in returns may reflect not only real differences but also inaccuracies or biased enumeration or estimation of the three projection bases for Ohio and the nation. First, the projection bases are examined for real differences in relative rates of growth and, second, they are assessed for statistical biases.

Estimation of Decade Totals, 1860-90

CHART 8

**Decennial Per Cent Change in Urban Population,
Ohio and the United States,
1840-1910**

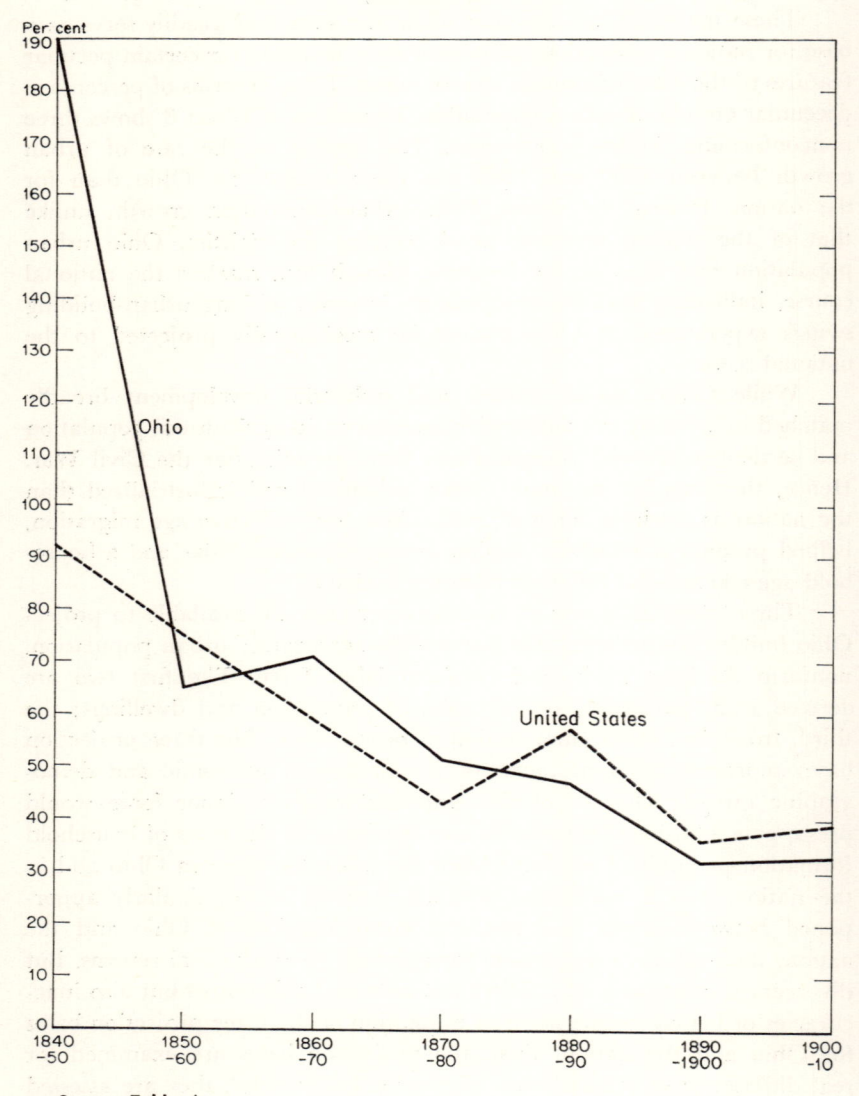

Source: Table 4.

Estimation of Decade Totals, 1860-90

The relative rates of dwelling-unit increment per unit of labor-force and urban-population increments in Ohio and the nation are presented in Table 8, along with actual nationwide increments in occupied nonfarm dwellings and national increments projected on Ohio rates (lines 15-19). Divergences between nonfarm dwellings and nonfarm labor-force increments were largely offset in the aggregate. For the first four decades, projection of nationwide housing increments on Ohio rates yielded figures smaller than those with the shortage concentrated in the first decade. During the 1900's, Ohio rates generated considerably larger figures than were realized nationally. In part, the decade variations reflect real divergences between Ohio and nationwide rates of household formation per unit of nonfarm labor force; in part, they reflect statistical inaccuracies in our nationwide measures of agricultural housing and labor. Enumeration of the nonfarm labor force in the U. S. Census was uncertain for agricultural workers, sometimes included in the category of general or unclassified workers. The Ohio enumeration may have been different from the national especially for the southern states with their fluid institutional patterns of farm operation.[28]

Another term in our comparison, occupied dwellings, was also difficult to adjust for farm dwellings by methods that could be applied uniformly in Ohio and the nation. Increments in farm dwellings can be gauged chiefly by increments in either farm establishments or farm labor force. The first measure is biased in the nationwide count by the breakup of plantations in the South, offset only in part by the rise elsewhere of large agricultural production units. Hence, the number of farms expanded after 1870 at a faster rate than farm families or farm dwellings did. On the other hand, hired farm labor was not consistently classified from census to census or possibly even for Ohio and the nation.[29]

The variations shown in Table 8, lines 13 and 14, between Ohio and nationwide rates of household increments per unit of urban population are much greater than the corresponding measure per unit of nonfarm labor force. The urban measure yields sizable underestimates for the first two decades. The cumulative tendency to underestimation was generated by higher rates of household formation per unit of urban

---

[28] From 1860 to 1910, farm labor force grew from 6,207.6 to 11,591.8, or by 86.7 per cent. Farms by number grew from 2,044 to 6,406, or by 213.4 per cent. *Historical Statistics,* 1949, Series D37, p. 72, Series K1, p. 278.

[29] For the nation as a whole, farm laborers accounted for 51.5 per cent of the agricultural labor force in 1860 and 39.2 per cent in 1910. Ohio shares were about half as large.

Estimation of Decade Totals, 1860-90

TABLE 8

Net Household Formation by Nonfarm Labor-Force Increment and by Urban-Population Increment, Ohio and the United States, by Decade, 1860-1910
(units in thousands)

	Increment	1860's	1870's	1880's	1890's	1900's
1.	Nonfarm labor-force increment					
2.	Ohio	106.0	153.0	276.3	258.7	368.3
3.	United States	1,525.4	3,016.7	4,491.5	4,521.7	6,892.9
4.	Nonfarm dwelling-unit increment					
5.	Ohio	53.9	39.8	129.5	111.9	171.8
6.	United States	952.1	901.0	2,110.5	2,023.8	2,941.7
7.	Urban-population increment					
8.	Ohio	282.5	347.8	479.4	488.2	666.8
9.	United States	3,685	4,228	7,976	8,054	11,839
10.	Dwelling units as per cent of increment Nonfarm labor force					
11.	Ohio (line 5/2)	50.86	26.99	46.78	43.25	46.65
12.	United States (line 6/3)	62.42	29.87	46.99	44.76	42.68
	Urban-population increment					
13.	Ohio (line 5/8)	19.08	11.44	27.02	22.93	25.76
14.	United States (line 6/9)	25.84	21.31	26.46	25.13	24.85
15.	National dwelling increment projected on Ohio rates					
16.	Labor force (line 11x3)	776	814	2,105	1,956	3,216
17.	Per cent of actual (line 16/6)	81.5	90.3	99.8	96.7	109.4
18.	Urban population (line 9x13)	703	484	2,155	1,847	3,050
19.	Per cent of actual (line 18/6)	73.8	53.7	102.1	91.3	103.7

Source: Tables 9 and 10; Eleventh Census, 1890, *Population*, Part I, p. 913; Thirteenth Census, 1910, *Population*, Part I, p. 1287. Line 5 derived from increment of total occupied dwellings less the farm increment of dwellings.

population in Ohio than in the nation. The average ages of Ohio urban population over the decades and the whole sweep of its age profiles were higher than those in the nation. More crucially, too, rates of urbanization per unit of growth in the nonfarm rural population were higher in Ohio than in the nation. New villages arose more slowly, and existing villages were more rapidly converted to urban status in Ohio than in the nation. Hence, rates of nonfarm housing growth per unit of urban-population increment would be lower in Ohio than in the nation. For these reasons, urban population was not used as a projection basis in our final calculations.

Granted that the statistical biases of the other two projection bases are less serious, are they satisfactory in other respects? Would rates of new residential building in Ohio per unit of our projection bases be

Estimation of Decade Totals, 1860-90

equal to nationwide rates? If so, it would imply that demolition and shrinkage rates were the same in Ohio and the nation, and that the shares of replacement construction to net growth were the same. Substantially, this is indicated by the scrappy evidence available. The relative share of the more durable brick and masonry structures was nearly the same in Ohio (13.3 per cent) as in the nation (13.9 per cent), to judge by the proportions of standing residential stock of those types in 1940.[30] The average age of housing stock in Ohio was probably somewhat under the national average age in the earlier decades of the projection, but continuous western settlement balanced off the newer regions against the older seaboard regions and probably brought the Ohio average dwelling age—and hence rates of shrinkage—into rough correspondence with nationwide levels. Such correspondence is indicated, at least, by the last quarter of the nineteenth century. The existing national shrinkage estimate for the three decades before 1920, stated in terms of per cent of dwelling production, amounted to 11.5, 8.2 and 7.1 per cent, respectively. Residential shrinkage rates in Ohio on the basis of annual counts between 1873 and 1884 probably approached 4 per cent in the 1870's and 5.5-6.0 per cent in the 1880's (see below, page 000). At any rate, the divergence between shrinkage rates, if these fragmentary data are to be trusted, would affect only a small fraction of total production. Hence, projection of Ohio replacement rates upon nationwide totals represents a fairly serviceable expedient, considering our needs, our objectives, and our margin of accuracy.

The work of projecting Ohio building experience into nationwide totals resolved itself into laying out three basic sets of decade aggregates: (1) Ohio and nationwide increments of nonfarm dwellings; (2) Ohio and nationwide increments of nonfarm labor force; and (3) new Ohio residential nonfarm production. In each category, estimation was involved in fixing the farm-nonfarm boundary. For dwellings, it could be gauged in the decades after 1890. Comparison of increments in farm establishments by number of establishments and number of farm families disclosed that farm-family increments were uniformly 25 per cent greater than farm establishment increments. The shrinkage factor was applied to farm establishment increments before 1890. The details of the adjustment are set forth in Table 9. Estimation of the nonfarm boundary in the labor-force increments was confined chiefly to adjustments in the nationwide

[30] *Sixteenth Census of the United States*, 1940, *Reports on Housing*, Vol. II, *General Characteristics of Housing, by States*, GPO, 1944, p. 68.

TABLE 9

CALCULATION OF ESTIMATES, NATIONAL NONFARM DWELLING INCREMENTS, 1860-1910

Estimation of Decade Totals, 1860-90

(units in thousands)

Census Year	Decade	Dwelling Stock Census Recorded[a]	Dwelling Stock Increment	Farm Stock[b]	Actual Increment	Estimated Increment	Estimated Increment, Nonfarm Dwellings
1860		5,628.6[c]		2,044.1			
1870	1860's	7,042.8[d]	1,414.1	2,660.0	616	462[e]	952.1
1880	1870's	8,955.8	1,913.0	4,008.9	1,349	1,012[e]	901.0
1890	1880's	11,483.3	2,527.5	4,564.6	556	417[e]	2,110.5
1900	1890's	14,430.1	2,946.8	5,737.4	1,172	923[f]	2,023.8
1910	1900's	17,805.8	3,375.7	6,361.5	625	434[f]	2,941.7

[a] 1850-80: 11th Census, 1890, *Population*, Part I, p. 913. 1890-1910: 13th Census, 1910, *Population*, Part I, p. 1287.
[b] from *Historical Statistics*, 1949, p. 95, E1-5.
[c] Adjusted upward on basis of 1870 Census report to include slave dwellings comparable with Negro dwellings of later census years.
[d] Not adjusted for undercount, which primarily involved agricultural population, not covered by the farm-dwelling adjustment.
[e] 75 per cent of value of actual increment.
[f] From Census record of increments of farm families, approximately equal to 75 per cent of farm increment. *Historical Statistics*, 1949, Series H89, p. 174.

Estimation of Decade Totals, 1860-90

count, designed to allow for the nonwhite-slave components of the southern labor force in 1860 and for the undercount of Negro labor in the South in 1870, as given in Table 10.

Estimation of Ohio residential dwellings erected were confined in the main to computed allowance for the share in this total of farm dwellings, set forth in Table 11. The procedure was to utilize decade increments in number of farms (column 2) as an acceptable measure of net change in farm dwellings, with some allowance for replacement production of farm dwellings. A special study of the five least-urban Ohio counties during 1900-10 indicated considerable farm residential-replacement building to cover losses from fire, demolition, or scrapping of older buildings.[31] Accordingly, we presupposed that statewide replacement rates, as set forth in Table 11, column 4, applied to farm dwelling stocks and calculated an allowance for gross farm dwelling production (column 6) and a parallel schedule of nonfarm residential production (column 8).

The details of the projection into nationwide housing estimates of Ohio nonfarm building rates, per unit of nonfarm-labor force and nonfarm-dwelling increments, are set forth in Table 12. All decade turns in nationwide dwelling production (Table 12, and Chart 9) are shared. Since the two multipliers yielded estimates close in pattern and level, and many of the biases inherent in each should be offsetting,[32] it seemed sensible to average the two estimates for projection purposes. The resulting level of decade estimates was adjusted upward slightly to fit the previously accepted estimates of 1900-20 (Table 12, line 7, and Chart 9).

The reliability of these decade estimates has been subjected to four tests wholly or partly independent: the first checks for decade pattern and level between 1860 and 1910; the second and third check over-all levels of building between 1860 and 1940; the fourth checks the decade of the eighties.

[31] The five counties had a 1910 percentage of urbanization ranging between 15.7 and 20.8. They contained, in 1910, 30,591 families and 13,028 farms. Over the decade, total population fell by 3,452; urban population rose by 4,214; while nonfarm, nonurban population fell by 3,835. The number of farms declined by 866. Both farms and rural dwellings of 1900 shifted categories over the decade. Yet, over the decade, 6,751 new dwellings were erected. Replacement building on farms is also indicated by the reported production of new barns and stables (3,093) and the increment (partly induced by price inflation) of value of building on farms ($8.2 million) averaging $632 per farm.

[32] Biases include: (1) possible nonuniformity by states in the administration of census definitions and enumeration procedures; (2) allowance for farm dwellings by 25 per cent rule (see p. 35); (3) delineation of agricultural laborer from "general laborer"; and (4) variation in the handling between Ohio and nationwide or between successive censuses of semiprivate households.

Estimation of Decade Totals, 1860-90

TABLE 10

Labor Force, Farm and Nonfarm, United States and Ohio, by Decade, 1860-1910

	1860 (1)	1870 (2)	1880 (3)	1890 (4)	1900 (5)	1910 (6)
		LABOR FORCE RECORDED				
United States						
All	9,425,133[a]	12,924,951	17,392,099	23,318,183	29,073,233	38,167,336
Farm	4,288,984[b]	6,263,394	7,713,875	9,148,448	10,381,765	12,582,997
Nonfarm	5,136,149[c]	6,661,557	9,678,224	14,169,735	18,691,468	25,584,339
Ohio						
All	640,043	840,889	994,475	1,287,101	1,545,952	1,919,055
Farm	302,798	397,613	398,188	414,544	414,662	419,423
Nonfarm	337,245	443,276	596,287	872,557	1,131,290	1,499,632

	1860's	1870's	1880's	1890's	1900's
	DECADE INCREMENTS OF RECORDED LABOR FORCE				
United States					
All	3,499,818	4,467,148	5,926,084	5,755,050	9,094,103
Nonfarm	1,525,408	3,016,667	4,491,511	4,521,733	6,892,871
Ohio					
All	200,846	153,586	292,626	258,851	373,103
Nonfarm	106,031	153,011	276,270	258,733	368,342

Source, by Column

(1) *Eighth Census, 1860 Population,* p. 399, pp. 656-680, for only free persons over 15 years old. Number of students was subtracted from Census figure.

(2) *Ninth Census, 1870, Compendium,* p. 594, corrected for understatement of 1870 Census. See *Sixteenth Census, 1940, Comparative Occupation Statistics for the United States, 1870 to 1940,* p. 104.

(3)–(5) *Twelfth Census, 1900, Occupations at the Twelfth Census,* pp. L and XCII.

(6) *Thirteenth Census, 1910, Classified Index to Occupations,* p. 41 and *Population,* Vol. 4, p. 125.

Notes: Technical Explanation of 1860 Adjustment

[a] Correction for 1860 Census was made as follows:
1. Number of students was subtracted from census figures.
2. To count slaves, we took number of slaves by sex, 15 years and over: male, 1,082,563; female, 1,074,748 (*Eighth Census,* p. 595). The number of gainfully occupied slaves was estimated by multiplying the above figures by 0.75 for male, 0.35 for female. These figures were added to the revised census figure 1.

Justification of choosing 0.75 and 0.35 as multipliers: For the years 1890, 1900, and 1910, we have 0.79, 0.84, and 0.87 for males and 0.36, 0.41, and 0.55 for females, 10 years-and-over age class. (*Negro Population in the U.S. 1790-1915,* Bureau of Census, 1918, p. 504.)

Estimation of Decade Totals, 1860-90

The percentage might be much higher for slaves in 1860, 15 years-and-over age class. But, for our purpose, it is necessary to estimate the total number of gainfully occupied persons in the same terms, because Ohio had no slaves in that year. Therefore, projecting back from the figures of 1890, 1900 and 1910, and taking account of difference in age class, the figures 0.75 and 0.35 were chosen.

b Col. 1, line 1 minus col. 1, line 3.

c To estimate the nonfarm slave population, the following steps were taken:

1. We estimated the number of gainfully occupied urban slaves by multiplying the total estimated number of slaves in col. 1 by 0.067, which was the estimated fraction of the total slave population living in urban communities (Gunnar Myrdal, *An American Dilemma*, New York, 1944, p. 183).

2. We estimated gainfully occupied, nonfarm slaves by multiplying the estimated gainfully occupied urban population by 1.29, calculated as follows: the nationwide figures of urban and rural population, and persons in agricultural pursuits for the years 1910 and 1860, and farm population for 1910 were taken from *Historical Statistics*, 1960. Then nonfarm, nonurban population for 1910 was calculated by subtracting farm population from rural population. To get farm population for 1860, the number of persons argiculturally employed in 1860 was multiplied by ratio of farm population to number of persons agriculturally employed in 1910. Nonfarm, nonurban population for 1860 was calculated by subtraction.

The ratio of nonfarm, nonurban population to the urban population was calculated—1.29. The underlying assumption is that the ratio of nonfarm, nonurban gainfully occupied persons to urban gainfully occupied persons is the same as that of population.

Finally, the total number of nonagriculturally occupied persons was calculated by adding the two figures, estimated above, urban slaves gainfully occupied and nonfarm, nonurban slaves gainfully occupied, to the census figure.

1. The first test reduces to an attempt to build a valid set of estimates of new residential production from the nationwide census returns of occupied dwellings by adjusting for uniformity and allowing for vacancy and replacement production. The details of the estimate are presented in Table 13. The first column presents the census returns, which were adjusted for 1850 to 1860 to include slave dwellings to ensure comparability with post-1860 returns. The second column presents an estimated set of vacancy allowances. For the years 1900 and 1910 we use Chawner's vacancy estimates, which appear reasonable on the surface having emerged from an exceptionally careful study of dwelling counts and family units.[33] Vacancy rates for other census years were gauged to the

[33] Chawner's vacancy estimates compare as follows with Wickens' (per cent):

	Chawner	Wickens
1900	2.53	4.0
1910	3.49	5.0

(Chawner, *Residential Building Process*, p. 16; Wickens, *Residential Real Estate*, pp. 54 ff.). The only nationwide time series of vacancy rates are those reported by the decennial English censuses, 1801-1911. The successive high-low variations run to less than two percentage points and, considering the greater age of the

Estimation of Decade Totals, 1860-90

TABLE 11

DERIVATION OF OHIO NONFARM HOUSING ESTIMATES, BY DECADE, 1860-1910

Decade	Statewide Reported New Dwellings (1)	*Farm Dwelling Adjustment* Increment Farms (2)	Farm Stock, Beginning of Decade (000's) (3)	Statewide Dwelling, Estimated Replacement Rate per 100 (4)	Estimated Replacement of Farms (3)×(4) (5)	Total Farm Adjustment (2)+(5) (6)	Replacement Increment of New Production (000's) (7)	Estimated Nonfarm Dwellings Produced (1)−(6) (8)
1860's	89,490	16,064	179.9	4.58	8,239	24,303	19.5	65,187
1870's	126,365	51,236	196.0	6.73	13,191	64,427	33.4	61,938
1880's	181,390	4,241	247.2	8.13	20,097	24,338	47.7	157,052
1890's	177,196	25,289	251.4	5.55	13,953	39,242	40.0	137,954
1900's	254,120	−4,674	276.7	10.13	28,030	23,356	86.9	230,764

SOURCE BY COLUMN

(1) From audited reports, adjusted for deficient returns, of county auditors, Ohio Secretary of State Reports.
(3) Bureau of Census, 13th Census, *Agriculture*, Vol. V, p. 68.
(4) Derived by dividing replacement increment of new production (col. 7) by the beginning decade total dwelling stock (from Census of Dwellings, *Compendium*, 1880, 1910).
(7) Derived by adding the increment of farms (col. 2) plus nonfarm dwelling units increment (Table 8, line 5) and subtracting the total from statewide reported new dwellings (col. 1).

40

Estimation of Decade Totals, 1860-90

1900-10 level, taking account of the position of the census year in the building cycle and the amplitude of movement of that cycle. The resulting vacancy estimates have little value in their own right and are used here only to adjust order of magnitudes. The determination of farm dwellings involved the adjustment of farm increments (see Table 9) to allow for the breakup of the plantation system.

Allowance for realistic estimates of conversion and shrinkage present complex estimation problems. The 1940 vintage report, showing that only 10.7 per cent of all dwelling units were converted (and this after a decade in which conversion activity was concentrated), indicates minimal decade rates of conversion, which were comparatively light in the decades between 1860 and 1910. Grebler, Blank, and Winnick have already accepted decade conversion allowances for later decades of the following percentages:[34] 1920's, 1.8; 1910's, 2.9; 1900's, 2.2; 1890's, 2.1. Accepting the indicated level, we accordingly specified the following pattern of conversion allowances, expressed as a percentage of dwelling production. A downward trend is specified to allow for the factor of aging:[35] 1900's, 2.2; 1890's, 2.0; 1880's, 1.8; 1870's, 1.6; 1860's, 1.4; 1850's, 1.2. The shrinkage ratio is more variable and involves larger magnitudes than conversion rates do. The BLS-NBER shrinkage estimates, drawn in large part from informed guesses by Wickens, were as follows (stated in terms of per cent of dwelling production):[36] 1930's, 15.0; 1920's, 8.3; 1910's, 11.5; 1900's, 8.2; 1890's, 7.1.

English dwelling stocks and higher tenancy rates, the absolute levels are comparable with those estimated by Chawner (see the British rates as reproduced in H. W. Robinson, *The Economics of Building,* London, 1939, p. 106). The Chawner estimates are also consistent with the carefully worked out 5.0 per cent estimate of Wickens for 1930 (based on an extensive analysis of 1930 and 1934 data, see pp. 22 ff.) and the census reported gross vacancy rates of 6.5 per cent for 1940 and 6.8 per cent in 1950, including vacant seasonal units and units held for absent households (see BLS-NBER, p. 776).

[34] *Capital Formation in Residential Real Estate,* p. 329 (drawing largely upon estimates by Wickens and Chawner).

[35] The whole scale of conversion allowances from 1850 to 1940 is too low, since it accounts for only little more than half of the total stock of 3.3 million conversion reported in the 1940 census count. But with present information there is no way to distribute the shortage over the decades; the tendency to conversion probably did not follow a uniform course. Hence, minimal conversion rates in line with the Grebler, Blank, and Winnick estimates are accepted for the purposes of the above calculation.

[36] See *Capital Formation in Residential Real Estate,* p. 329.

Estimation of Decade Totals, 1860-90

TABLE 12

DERIVATION OF UNITED STATES HOUSING ESTIMATES BY USE OF
OHIO MULTIPLIERS, BY DECADE, 1860-1910
(thousands)

Ohio Multipliers	1860's	1870's	1880's	1890's	1900's	Total
1. Nonfarm dwellings produced in Ohio per 1,000 nonfarm occupied dwelling increments (Table 11, col. 8, Table 8, line 5)	1,208.7	1,557.8	1,212.8	1,232.5	1,342.9	
2. Nonfarm dwellings produced per 1,000 nonfarm labor force increments (Table 11, col. 8, Table 8, line 2)	615.0	404.8	568.4	533.2	626.5	
3. Variant estimates, nationwide dwellings production						
4. Line 1, above, Table 8, line 6	1,150.8	1,403.5	2,559.6	2,494.3	3,950.5	11,558.4
5. Line 2, above, Table 8, line 3	938.1	1,221.2	2,552.9	2,410.9	4,318.3	11,441.4
6. Average	1,044.5	1,312.4	2,556.3	2,452.6	4,134.4	11,499.9
7. Average adjusted to Chawner 1900-19 level[a]	1,061	1,333	2,597	2,491	4,200	11,682

[a] Upward adjustment of 101.58 per cent.

The Ohio report of buildings lost, destroyed, or demolished showed the following ratios to reported production of new dwellings in the same year.

Year	Number	Value	Year	Number	Value
1873	3.60	7.76	1878	4.32	4.62
1874	3.60	7.93	1880	4.69	5.97
1875	3.96	8.51	1881	5.25	14.40
1876	3.64	9.00	1882	5.40	6.52
1877	6.44	7.67	1884	6.30	12.40

For residential building, only the number estimates would be relevant. It would be expected that in the eastern states higher shrinkage

Estimation of Decade Totals, 1860-90

CHART 9

United States Decade Totals for New Dwellings, Three Variants, 1860-1910

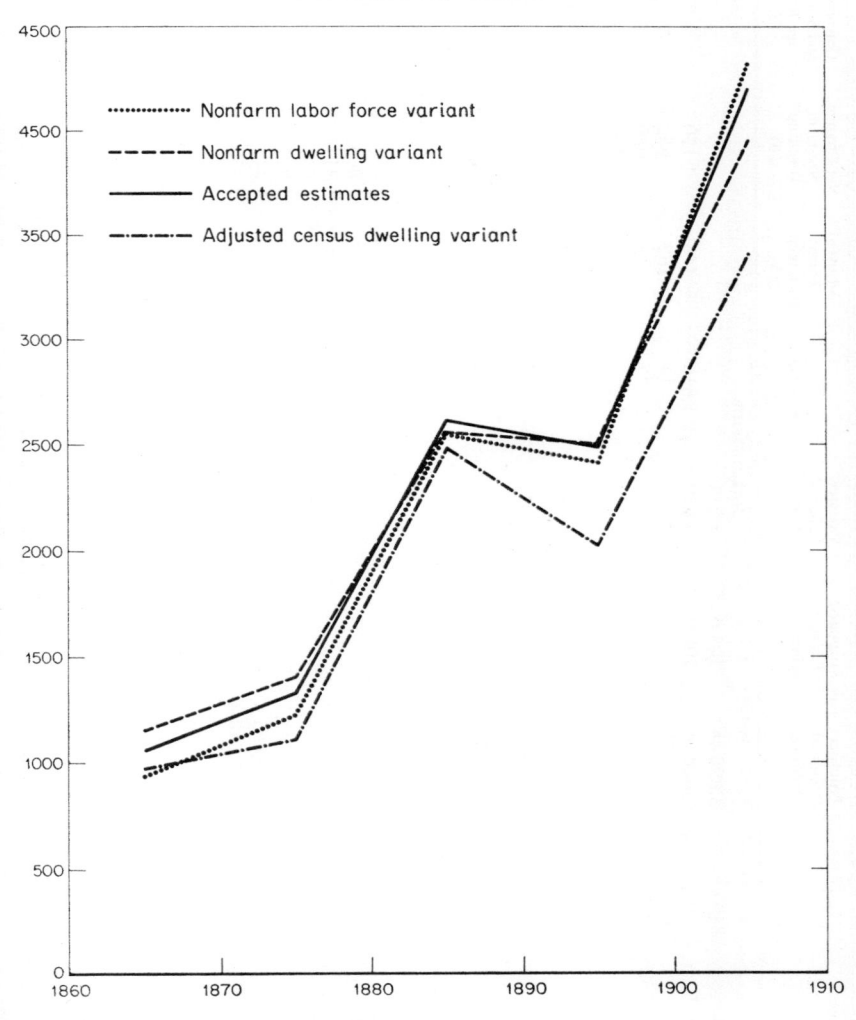

[Chart showing four variants from 1860 to 1910:
- Nonfarm labor force variant
- Nonfarm dwelling variant
- Accepted estimates
- Adjusted census dwelling variant]

Source: Table 12, lines 4, 5, 7; Table 13, col. 8.

Estimation of Decade Totals, 1860-90

TABLE 13
Calculation of Adjusted Housing Production from Census Dwelling Increments, 1860-1910

Census Year	Decade	Number Occupied Dwellings (1)	Estimated Vacancy Allowance (2)	Vacancy Adjusted Stock (3)	Increment (4)	Unadjusted Farm Increment (5)	Adjusted Farm Increment (5)	Net Change, Nonfarm Dwelling (6)	Conversion and Shrinkage Rate (7)	Estimated New Gross Production (8)
				NATIONWIDE						
1850		3,896.4[a]	103.0	4,013						
1860	1840's	5,628.6[b]	100.0	5,629	1,616		446	1,170	1.8	1,191
1870	1850's	7,042.8[c]	100.0	7,043	1,414		462	952	2.3	974
1880	1860's	8,955.8	102.0	9,135	2,092		1,012	1,080	2.9	1,112
1890	1870's	11,483.3	104.0	11,942	2,807		417	2,390	3.7	2,482
1900	1880's	14,430.1	102.5	14,791	2,849		923	1,926	4.7	2,021
1910	1890's	17,805.8	103.5	18,429	3,638		434	3,204	6.0	3,409
	1900's									9,998[d]

Census Year	Decade	Occupied Dwelling Stock (9)	Vacancy-Adjusted Dwelling Stock (10)	Unadjusted Farm Increment (11)	Estimated New Gross Production After Conversion and Shrinkage (12)
			NATION EXCEPT SOUTH		
1850		2,343.5	2,413.8		
1860	1840's	3,645.2	3,645.2	437.7	808
1870	1850's	4,742.5	4,742.5	403.1	711
1880	1860's	5,948.7	6,067.6	702.9	641
1890	1870's	7,849.7	8,163.7	250.4	1,917
1900	1880's	9,685.6	9,927.8	388.7	1,443
1910	1890's	11,868.2	12,283.6	147.0	2,350
	1900's				7,062

SOURCE BY COLUMN

(1) & (9) 11th and 13th Censuses, *Population*, Part I, pp. 913, 1287, respectively.
(2) See text, pp. 39ff.
(5) Table 9.
(7) Determined by subtracting estimated conversion rates (p. 41) from loss rates (p. 45).
(11) Actual count of number of farms, first difference. *Abstract of Census*, 13th Census, 1910, p. 283.
[a] Adjusted to include slave dwellings on par with post-1860 census returns; calculated at rate of six slaves per dwelling (1860 rates).
[b] Adjusted by 1870 Census report to include slave dwellings comparable with post-1860 census returns.
[c] Not adjusted upward for 3.3 per cent population undercount, which did not seem to have affected the farm count or the nonagricultural sector.

Estimation of Decade Totals, 1860-90

rates would prevail. A rising tendency is indicated, though in part it may reflect the cyclical movement in process. Other reports have shown that shrinkage activity and demolition accelerates in the building boom phase.[37] The years of the early eighties involved a marked boom. In selecting shrinkage rates, consideration was also given to reported shrinkage rates elsewhere involving average annual replacement as a per cent of new residential production:[38] Canada, 1920-1940, 19.4; Hamburg, 1885-1913, 13.2; Glasgow, 1873-1913, 45.0. The Glasgow rates as compared with the Hamburg rates disclose the influence of high average age of stock and slow rates of secular growth. Allowance must also be made for the temporary character of nonfarm dwellings in the first round of settlement, the tendency to quick obsolescence, and relatively high rates of fire loss. In the later decades, demolition would accelerate in metropolitan communities owing to programs of highway and road building and revamping of city layout and design. In view of the Ohio demolition report, it seemed safe to commence with a relatively low 3 per cent rate for the 1850's. The progression ends with the comparatively well-established rate of 15 per cent for the decade of the 1930's. Intermediary rates were logarithmically interpolated.

The crudeness of this procedure is recognized. Loss rates on total housing stock are probably a function of its average age, percentage share of wood to masonry construction, and rate of new residential building. There is no reason this interplay of influence should work out to a steady logarithmic progression of loss rates expressed as a fraction of new decade residential production. However, in the absence of intensive research on loss rates and with uncertainty regarding absolute decade loss levels, we have accepted the following percentages for the United States, 1850-1940: 1850's, 3.0; 1860's, 3.7; 1870's, 4.5; 1880's, 5.5; 1890's, 6.7; 1900's, 8.2; 1910's, 10.0; 1920's, 12.3; 1930's, 15.0.

Table 13 also presents a similarly derived schedule of estimated gross residential production for the entire nation except the South. Since

---

[37] Thus Ohio decade replacement rates (see Table 11) were 8.13, 5.55, and 10.13 per hundred dwellings of stock for the eighties, nineties, and nineteen-hundreds.

[38] See A. K. Cairncross, *Home and Foreign Investment 1870-1913*, London, Cambridge University Press, 1953, p. 26; K. Hunscha, *Die Dynamik des Baumarkt*, Vierteljahreshefte Zur Konjunkturforschung, Sonderheft 17, Berlin, 1930, p. 60; O. J. Firestone, *Residential Real Estate in Canada*, University of Toronto Press, 1951, pp. 382 ff., 393 ff. Chawner used demolition rates for the period 1900-1936 ranging from 2.5 units per 10,000 persons in 1926 and 1928 to 6.0 units in 1935 (*Residential Building Process*, p. 14).

Estimation of Decade Totals, 1860-90

nonsouthern farm increments corresponded more closely with the indicated real movement of agricultural population and its dwelling stocks, the farm-dwelling increments were unadjusted. This leads to a probably excessive spread between the national and nonsouthern or to an exaggerated account of southern nonfarm dwelling production. Our estimates presuppose that the nonsouthern part of the nation contributed only 71 per cent of the nonfarm residential production between 1860 and 1910. The share, judging by measures of change in urban population of nonfarm labor force, may well approach 80 per cent. A more precise regional breakdown of nonfarm residential building was not feasible within the limits of this study.

The difficulty in reconciling the regional accounts underscores the crudeness of our efforts to build up from census returns an estimate of new residential production. The aggregate level estimated falls short by 13 per cent of the level projected by the two Ohio multipliers, which we have accepted. In part the shortage derives from our inability to adjust for underestimation in the census dwelling count between 1850-1910.[39] More important is the decade pattern of movement as shown in Chart 9, which conforms closely to the pattern of our accepted estimates.

2. A second test of these estimates provides a check on the aggregate level of residential building involved—11.7 million new nonfarm dwellings between 1860 and 1910. Between those years nonfarm population to be housed grew by approximately 45.6 million.[40] The average-size

---

[39] Thus for 1900 and 1910 we used an unadjusted census count of 14,430,100 and 17,805,800 occupied dwellings (see Table 13). The adjusted totals for households used in *Historical Statistics*, 1949, H89-112, were 110.0 and 113.8 per cent greater.

[40] For 1910, total nonfarm population was recorded by the Census at 59,895,200. The problem is thus to build up an 1860 nonfarm estimate. Of the two components of this estimate—urban and nonurban—the urban is recorded at 6,216,500. We build up the nonurban component by assuming that resident farm population will show the same pattern of movement as total agricultural employment (as taken from the occupations census). This is a more reliable basis for projection than movement in number of farms. Farm-family size patterns probably did not shift much over the fifty-year period. The numbers (in thousands) are as follows:

	1860	1910
Urban population	6,216.5	41,998.9
Rural population	25,226.8	49,973.3
Farm employment	6,207.6	11,591.8
Nonfarm, nonurban population	8,048.6 (derived)	17,896.3
Total nonfarm population	14,265.1	59,895.2

SOURCE: *Historical Statistics*, 1960, pp. 25-29, 63.

Estimation of Decade Totals, 1860-90

nonfarm household to be accommodated in 1910 was 4.24 persons,[41] which would call for 10.76 million dwellings. If we allow for a 3.5 per cent vacancy in the end stock in 1910 and if we specify a 6 per cent shrinkage (for fire and loss) in post-1860 production as consistent with the last tabulation, we emerge with a total new residential requirement over the fifty years of 11.8 million units.[42] This is close to the magnitude of our accepted decade levels.

3. A third check of over-all magnitudes is made possible by the Housing Census of 1940. Cumulation from 1860 to 1939 of all our estimates yields a total of 27.4 million dwellings. A comparison of that total (in thousands) with the known 1860 and end-1939 standing stock of nonconverted eligible nonfarm dwellings is shown in the next tabulation.

1. Estimated new production nonfarm units 1860-1939	27,418
2. Estimated 1860 nonfarm stock	3,167[a]
3. Total available (items 1 plus 2)	30,585
4. Surviving eligible end-1939 standing stock of nonconverted nonfarm units	26,211[b]
5. Gross disappearance (items 3 minus 4)	4,374
6. Disappearance as per cent of new production (item 5 as per cent of item 1)	15.95

[a] Average of two estimates: (a) using nonfarm population estimate for 1860 (see n. 40) and latest average nonfarm family size (4.24 persons), and (b) taking households, 1860 Census, minus "free" farms, plus estimated slave urban families (from Table 13 at 43,000).

[b] Total census standing stock less converted and "ineligible" units and less estimated transfer from the farm sector (Table 2, and 16th Census, *Housing*, 1940, III, Table A-1, p. 9.)

The number of traceable 1860 units reported in 1940 as surviving through to 1940 was 633,000 units. If that count, probably an understatement, be conceded, the disappearance rate of the post-1860 production was 6.71 per cent, a not implausible rate for post-1860 stock. The total disappearance rate, 15.95 per cent, is very close to that recorded for Canada between 1920 and 1940 (see footnote 38). It is worthwhile to compare this total disappearance rate with the disappearance rate implicit in the scale of loss rates developed by Wickens and by Grebler *et al.* for 1890-1940, and extended logarithmically and smoothed in our reference scale back to 1850 (tabulation, p. 45). Applying the reference

---

[41] This includes quasi-households as counted in the 1910 Census (see *Capital Formation in Residential Real Estate*, pp. 82, 393 ff.).

[42] Details are as follows (in thousands): vacancy rate of 3.5 per cent (Chawner's estimate), 377; 6 per cent shrinkage from current production (fire and loss), 646; estimated net dwelling requirement, 10,759; total required production gross, 11,782.

Estimation of Decade Totals, 1860-90

CHART 10

**Annual Mortgage and Building Activity,
Nationwide, 1880-89**

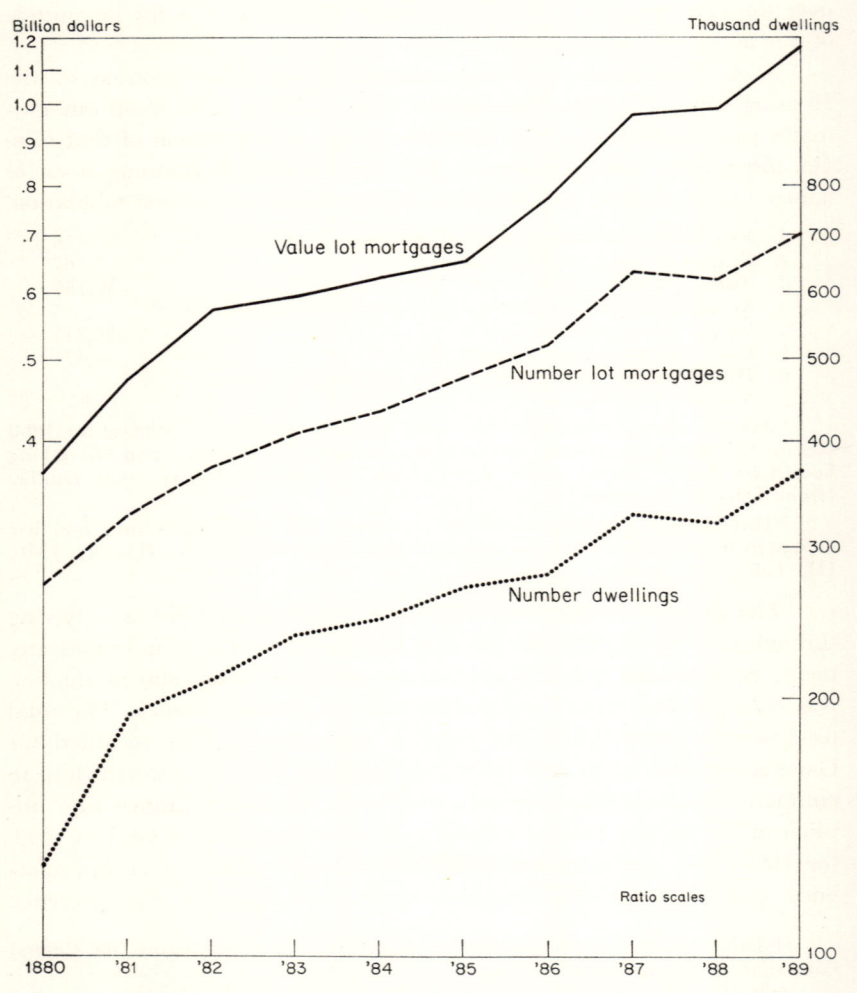

Source: U.S. Census, *Report on Real Estate Mortgages*, 1890, p. 317; and Table 15, below.

Estimation of Decade Totals, 1860-90

shrinkage scale yields a gross disappearance over the eighty-year period of only 2,709,000 units, appreciably less than gross disappearance implicit in our measures (4,374,000 units). I suspect the gap owes more to understatement of the reference shrinkage scale than to overstatement in our estimates of new production.

4. A fourth test is available for the decade of the 1880's. The 1890 Census included reports of a nationwide survey of the number and value of mortgage instruments recorded annually from 1880 to 1889 in connection with real estate located in municipalities of all size classes. Mortgage recordings should move coordinately with series of new residential building, though with dampened amplitudes.[43] Chart 10 plots the movement of our estimated national housing series and the census-derived nonfarm mortgage series. Patterns of change are congruent both in form and amplitude.

[43] Specific total average amplitude measurements for long building cycles for Ohio 1880-1900, are as follows (in thousands):

	Residential Building	Value of Mortgage Loan Recordings
Mean of 3 sample Ohio groups	207.0	140.8
Statewide aggregate	120.0	66.9

4
Estimation of Decade Totals, 1840-60

The backward projection for the 1840's and 1850's must be made without direct benefit of the Ohio bench-mark production. For the decade of the 1850's, two bases for projection are available giving quite different results. The nationwide census of dwellings, when adjusted to show nonfarm decade increments, shows an increment for the 1860's 18.2 per cent less than that for the 1850's (see Table 13). The entire country except the South, with its special problems of household and farm identification, shows a 12 per cent decline for the sixties. The census of occupations, however, shows that incremental growth of nonagricultural employment in the sixties was 14.4 per cent ahead of that in the fifties. The rate of growth of urban population manifested a similar rise. If the order of magnitude of these census counts and derived estimates is valid, it can only mean that dwelling production in the fifties ran ahead and in the sixties ran behind the underlying population growth. To this extent the building boom of the fifties must have corresponded in intensity with the building boom of the eighties, the profiles of which show up clearly in the behavior of Ohio rates of nonfarm residential building per unit of growth of urban population or nonagricultural employment.[44] For the fifties we accept the dwelling count for the country minus the South, adjusted for the count of farm increments, and record for the fifties a decade total of 113.6 per cent of that accepted for the sixties. Such a decade total is consistent with such annual index behavior as we have for

[44] See comparison of the following rates of dwelling production:

	Ohio Nonfarm Dwelling Units Produced per 1,000 Urban Population Increase	Estimated Nationwide Dwelling Units Produced per 1,000 Nonfarm Employment Increment
1860's	230.8	696
1870's	178.1	442
1880's	327.6	578
1890's	282.6	551

SOURCE: Tables 1, 8, and 11.

Estimation of Decade Totals, 1840-60

the decade; and the implicit rate of decade increase is free from problems of southern estimation.

The decade of the forties presents other problems. We have no dwelling count for it. The incremental growth of nonagricultural employment and of urban population (which was then the smaller portion of nonfarm population) was only 71.6 per cent and 63.5 per cent, respectively, of the corresponding magnitudes of the fifties. This suggests a smaller volume of dwelling production for the forties than for the fifties. But rates of dwelling production for these magnitudes can vary widely, as we have just seen. If the boom of the early and middle fifties was exuberant, the long-swing contraction of the early forties was equally prominent in the scattered records of the time. It would be improper to impress this judgment, however well supported it may seem, into our statistical series. At this juncture it is fortunate that the 1840 *Census of Manufacturers* included a count of dwellings erected in the census year of record. Nationwide, some 54,113 dwellings were enumerated. Recent use of the 1840 census returns indicates that its results hold up under careful scrutiny, but that adjustment is often needed for careless enumeration or for omissions.[45] The reliability of the census housing count was subjected to a twofold evaluation for Ohio and the nation.

The housing returns by counties in Ohio for number and value were expressed as number of houses built per capita and as value per house built, per person, and per dollar of assessed value of real property. Analysis of the value returns showed that returns for three relatively large counties involved probable valuation error because of abnormally high housing values per house relative to total assessed property.[46]

---

[45] See use of the 1840 Census return by Gallman, "Commodity Output, 1839-1899," and by Richard A. Easterlin, "Interregional Differences in Per Capita Income, Population, and Total Income, 1840-1950," *Trends in the American Economy*, Studies in Income and Wealth, Vol. 24, Princeton for NBER, 1960.

[46] The reported values of newly erected houses expressed per house built and as a fraction of total assessed real property in 1841 were as follows:

County	Dollar value per house	Value of houses as fraction of assessed real property
Franklin	$5,688	2.53%
Hamilton	3,188	13.88
Licking	4,389	27.61
Statewide (less above three counties)	522	1.97
Original statewide	1,012	3.81

Whereas the enumeration in Franklin County probably understated housing numbers (indicated by absence of returns for wooden houses), the enumeration for Hamilton and Licking clearly overstated values.

Elimination of these returns reduces the average value per house from $1,012 to $639, which compares closely with the nationwide census average of $775.[47] However, even this level of value was felt to be untrustworthy. Valuation errors must have been widespread since many—perhaps most—dwellings were erected wholly or largely by owner occupants using, in considerable measure, unpurchased materials. Ground values and investment in site improvements may also have been incorporated in estimates of value.

Number of houses is a more workable magnitude than housing values. The reliability of the number count was checked by comparing rates of per capita building against rates of population growth, in scatter diagrams of county observations for 1840 Census population per houses built, showing (1) the decennial rate of population change between 1830 and 1840, and (2) per cent growth of assessed real estate in 1841 over 1835. The relationship indicated by use of method (2) was more systematic (see Chart 11). Seventeen counties were excluded as they lacked the requisite 1835 valuation bench mark or record of housing production. As expected, we found a broad field of scatter on a semilog scale for the 62 correlated counties but a perceptible tendency of high rates of building to correlate positively with high rates of assessed property growth. The broad linear band excluded 18 counties as having improbable rates of per capita building in view of rates of indicated property growth. Two of the excluded counties specified the number of masonry houses but not of wooden, and a few others specified only wooden but no masonry construction. Ten of the excluded counties with enumerated population of 9 per cent of the state aggregate lacked any entry on housing enumeration, though economic indexes indicated growth over the decade and since 1835.[48] The housing census was taken as part of the schedule of manufacturing, which enumerators might readily neglect in newly settled western country. There are thus grounds for presuming that the housing count for the included 44 counties was more reliably recorded than that

---

[47] The computed $522 (see footnote 46) was adjusted to a statewide value allowing for the differential in value per house in the excluded three counties and rest of state in the middle 1860's.

[48] The ten counties experienced a population growth during 1830 of 38.6 per cent or considerably less than the state average. Four of the counties with little or no increase in cultivated acreage between 1835 and 1841 experienced a 50 per cent rise in the assessed value of town property. The other six counties experienced a market rice in both farm acreage and town property. Altogether, it is unlikely that an area holding 9 per cent of the state's population would be totally devoid of residential building during 1839-40.

Estimation of Decade Totals, 1840-60

CHART 11

Number of Persons Per House Erected in Ohio, by Counties, 1840, and Per Cent Growth of Assessed Real Property, 1841 Over 1835

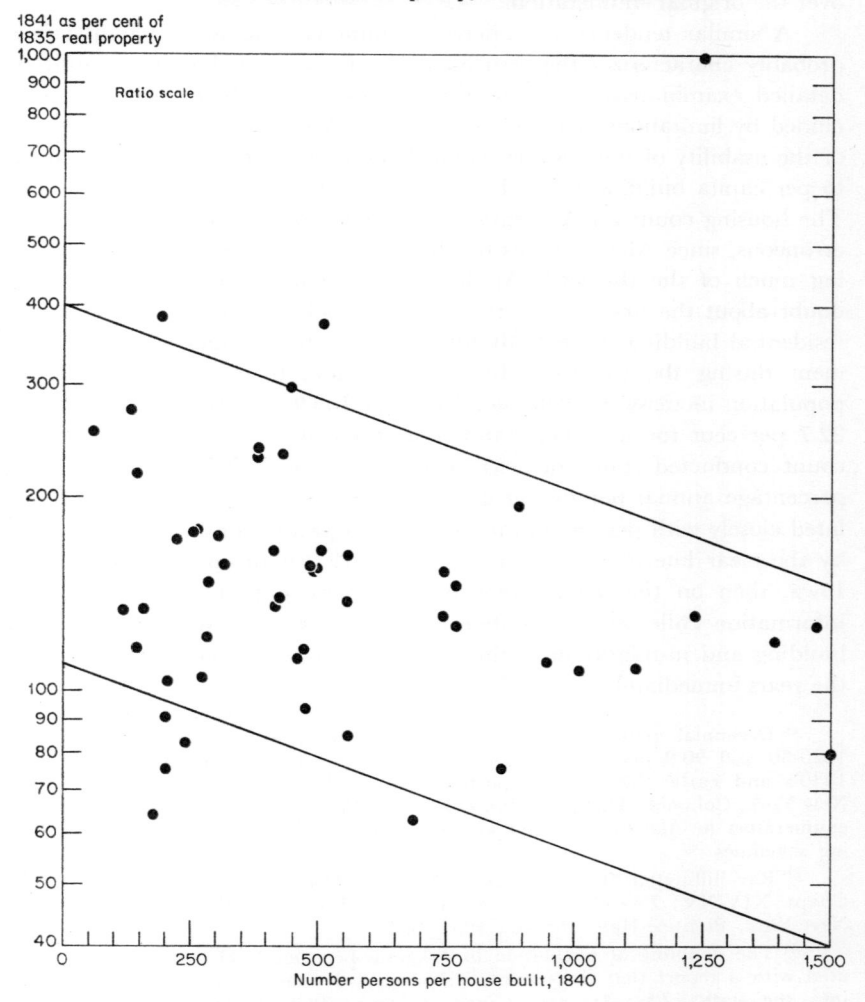

Note: Nine observations with high values not shown on this restricted scale.
Source: Ohio Board of Equalization, *Proceedings*, Columbus, Ohio, record of valuation by counties, 1835, pp. 28-30; abstract ... of valuation by counties, 1841, pp. 60-62; *Compendium of the 6th Census*, 1840, pp. 326ff; *Ohio Population*, p. 26.

for the excluded counties, and that it may serve as a bench mark for projecting statewide rates of building. If we apply the per capita rates of building for the included counties to the statewide aggregate, we derive a total statewide housing production of 4,159 units or some 11.4 per cent over the original enumeration.

A similar tendency to undercount numbers and to overcount value probably characterizes the returns of the nationwide housing census. A detailed examination of the county returns for all other states was precluded by limitations of time and resources. A check was made, however, of the usability of the various statewide housing counts by reducing them to per capita building rates. The array of rates is presented in Table 14. The housing count for Alabama, with its very low rate of building, seems erroneous, since Alabama was on the frontier of western settlement during much of the thirties.[49] At the other extreme there was ground for doubt about the first nine states with extremely high rates of per capita residential building in 1839-40, all located in the frontier zone of settlement during the thirties.[50] In the aggregate, their decennial rate of population increase in 1840 over 1830 was 169.9 per cent compared with 22.7 per cent for all other states. For seven of the states, a population count conducted statewide was available between 1835 and 1838. The percentage annual population growth in the years preceding 1840 correlated closely with per capita rates of housing production in 1840 as shown by the clear line of regression (see Chart 12), with only the return for Iowa, then on the outer fringe of settlement, out of line. Qualitative information collected by frontier historians also indicates that rates of building and in-migration in the frontier states were sometimes heavy in the years immediately before 1840.[51]

---

[49] Decennial population increases for Alabama were 142.1 per cent for 1820-30 and 90.9 per cent for 1830-40. Relatively heavy migration in the late 1830's and early 1840's was reported (Minnie Boyd, *Alabama in the Fifties*, New York, Columbia University Press, 1931, p. 18). It seems probable that Census enumerators in Alabama did not enumerate slave cabins under the manufacturing schedules.

[50] Ray Billington, *Westward Expansion*, New York, Macmillan, 2nd ed., 1960, Chaps. XIV-XV; *The Growth of the American Economy*, H. F. Williamson, ed., New York, Prentice-Hall, 2nd ed., 1951, p. 93.

[51] The doubling of population in Arkansas between 1835 and 1840 is associated with a report that the panic of 1837 caused many eastern farmers to pour into the state (*The Arkansas Guide*, Arkansas Writer's Project, WPA 1941, p. 39). The nationwide crisis of 1837 was registered in Arkansas in 1841 according to another report (*Arkansas Centennial*, Arkansas Democrat, 1936, p. 12). William Pooley, in his review of the settlement in Illinois between 1830 and 1850, dated the peak of settlement in the period 1837-40 and believed that much of it

Estimation of Decade Totals, 1840-60

TABLE 14

ARRAY OF PER CAPITA RESIDENTIAL BUILDING RATES,
TWENTY-NINE STATES, 1840

State	Number of Houses Built per 1,000 Persons
1. Wisconsin	16.64
2. Iowa	11.56
3. Arkansas	11.26
4. Illinois	9.38
5. Missouri	6.81
6. Indiana	6.73
7. Mississippi	6.36
8. Michigan	6.22
9. Florida	5.79
10. Georgia	3.80
11. Maine	3.40
12. Kentucky	2.87
13. South Carolina	2.87
14. New Jersey	2.86
15. Rhode Island	2.73
16. New York	2.64
17. Pennsylvania	2.57
18. North Carolina	2.47
19. Louisiana	2.46
20. Ohio	2.46
21. Virginia	2.42
22. Massachusetts	2.13
23. Maryland	2.09
24. Connecticut	1.97
25. Delaware	1.94
26. Vermont	1.85
27. New Hampshire	1.83
28. Tennessee	1.56
29. Alabama	0.91

SOURCE: *Compendium of the 6th Census,* 1840, p. 361; *Historical Statistics,* 1960, p. 13.

Estimation of Decade Totals, 1840-60

CHART 12

Scatter Diagram of Percentage Growth of Population, Annually, 1835-40, and Per Capita Housing Production, Seven States, 1840

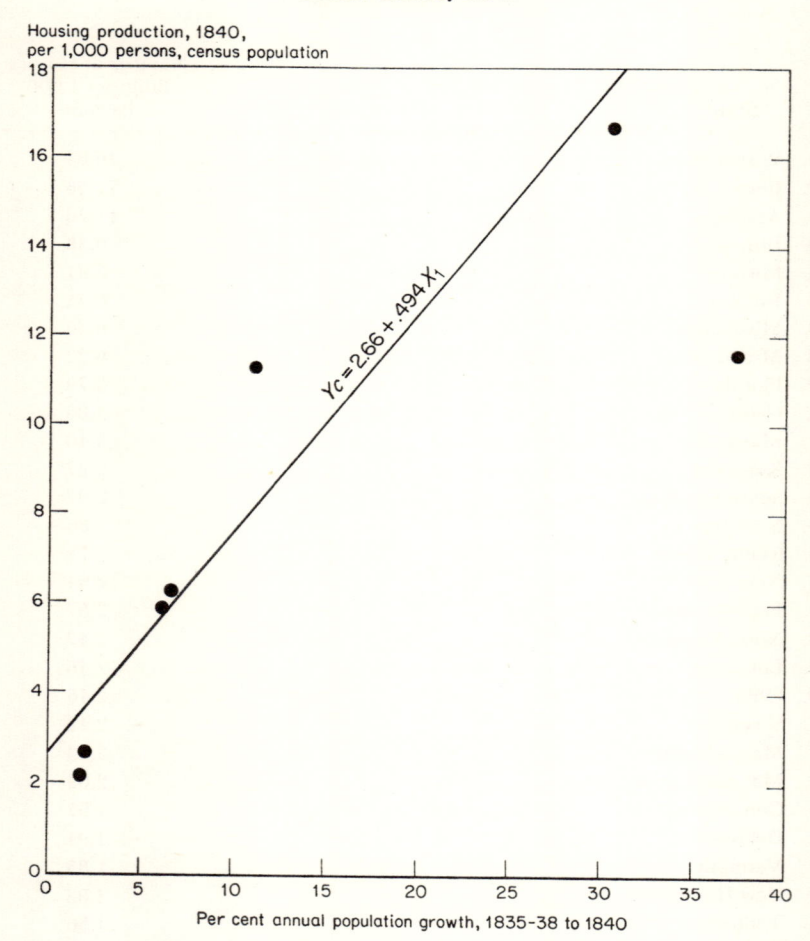

Housing production, 1840, per 1,000 persons, census population

$Y_c = 2.66 + .494 X_1$

Per cent annual population growth, 1835-38 to 1840

Source: Figures for 1840 from Table 14. Figures for 1835 or 1838 from: Wisconsin—John Hunt, *Wisconsin Almanac*, 1856, p. 79; Florida—*Florida Becomes a State*, Florida State Library, 1945, p. 132; Arkansas—*Arkansas Democrat, Arkansas Centennial*, 1936, p. 3; Iowa—*Iowa Guide*, Federal Writers Program, 1938, p. 550; Michigan and Massachusetts—Murray Kane, "Some Considerations on the Safety Valve Doctrine," *Mississippi Valley Historical Review*, Sept. 1936, pp. 183-4; New York—*Census of State of New York*, for 1835.

Estimation of Decade Totals, 1840-60

If returns throughout the range of the per capita building array are consistent, it is reasonable to hold that the number count of houses built in 1839-40 is acceptable, subject only to the qualification that adjustment would be needed for omitted counties or categories of building. Using our revised Ohio rates of per capita building as a guide, we estimated national production at 54,359 dwellings, or slightly over the enumerated quantity. Since rates of population growth in Ohio exceeded the national average, Ohio building rates per capita should not tend to understate. Regression analysis also adds confirmation to the recorded census count.[52] In any case our quantity needs to be adjusted by an allowance for farm dwellings erected in the census year. Here, too, all we know is the decennial distribution after 1840 of occupied labor force increments between agricultural and other pursuits. This distribution suggests that in any given year an appreciable share—from perhaps one-third to three-fourths —of the dwellings erected in 1839-40 must have been nonfarm. Are we far off in concluding that any estimate for 1840 residential nonfarm building falling within a range of 16,000 to 45,000 would lie within the range of acceptability indicated by the evidence brought under review? If we allow for this order of magnitude and accept the annual behavior of our available annual indexes and bear in mind the need for a fit with the succeeding decade, we find a total of 520,000 units for the forties

---

was in reaction to the 1837 crisis and crop failure in New England ("The Settlement of Illinois from 1830-1850," *University of Wisconsin Bulletin*, History Series, Vol. I, 1908, pp. 337 ff., 339 ff., 568 ff.). Murray Kane conducted a careful review of population and migration data during the thirties and concluded that the balance of evidence supported the hypothesis that the migration movement westward "reached a peak just prior to the precipitation of the panic and contracted during recession." He collected evidence that leading frontier urban centers, such as Cincinnati, Chicago, and Detroit, were affected by nationwide business cycles with little lag ("Some Considerations on the Safety Valve Doctrine," *Mississippi Valley Historical Review*, Sept. 1936). Kane took little reckoning of the fact that the upswing of the thirties was twin peaked and that the "recession" for many communities may easily date from 1839 or 1840. The rejoinders to Kane's work and other critical articles and the subsequent discussion provided only meager additional evidence (see review of literature in Billington, *Westward Expansion*, 2nd ed., 1960, pp. 762 ff.). The 1840 housing census, when analyzed in terms of the "safety valve" hypothesis, makes up a significant item of new evidence.

[52] If we use a freehand linear regression, and utilize the 2.57 per cent average annual rate of growth of nationwide population between 1837 and 1840 (for annual estimates see *Historical Statistics*, 1949, p. 26), we derive an estimated number of 3.3 houses built per 1,000 of 1840 census population; and this applied to the 1840 census population yields 56,497 dwellings. The least squares regression, $Y_c = 2.66 + .494X_1$, based on six observations, yielded a national estimate of 67,083,000 houses.

Estimation of Decade Totals, 1840-60

(43.3 per cent of the estimated total for the fifties) would be a plausible estimate. It would fit the 1840 order of magnitude, the annual time series, and the pattern of decade increments. A decade production on that scale, however, would have involved relatively low rates per unit of urban increment or nonfarm employment. The rates compared with the succeeding decades are shown in the next tabulation.

	Size of Increment (thousands)			Residential Production per 1,000 of Increment[a]		
	1840's	1850's	1860's	1840's	1850's	1860's
Nonfarm employment	1,095.3	1,528.8	1,750.1	474.8	786.2	606.2
Urban population	1,699	2,673	3,685	306.1	449.7	287.9

[a] Assuming dwelling production of 520,000 units in 1840's, 1,202,000 in 1850's, and 1,061,000 in 1860's (*Historical Statistics*, 1949, Series A195-209, p. 14; Series D38, p. 72).

I regard the stipulated 1840 rates of housing production as well within the range of possibility. Statewide quinquennial reports on building materials produced in Massachusetts show very high amplitudes of rise from 1840 to 1855.[53] And Gallman's quinquennial estimates for deflated value of total construction involve a pattern of movement on a scale comparable to that of our estimates.[54] For these reasons and with the indicated reservations, our decade projections and annual series include a set of estimates for the 1840's. These estimates are regarded as only showing the pattern of probable long-swing movement and as defining the outer limit of the long-swing amplitude consistent with all our present information.

---

[53] Massachusetts state census reports show the following per cent increases for 1855 over 1840 for dollar value at current prices: bricks produced, 845; building stone, 147; lumber planed or sawed, 1,062.

[54] With 1839-40 as a base, the quinquennial index relatives for four quinquennial years are appended:

Year	Our Residential Estimates	Gallman Construction, Deflated Total Value	
		Variant A	Variant B
1839-40	100	100	100
1844-45	128	117	121
1849-50	261	155	164
1854-55	383	281	293
1859-60	258	294	318

Source: Table 15, and Gallman, "Commodity Output, 1839-1899," *Trends in the Nineteenth Century*, p. 63, lines 6, 11.

Since residential building would be expected to fluctuate more violently in its growth path than total construction, the two sets of series are reasonably consistent, especially in over-all gradient.

5

Estimation of Decade Annual Indexes

We now turn to the third stage of the procedure used in this study, the fixing of suitable decade indexes, i.e., sets of ten index relatives of housing production in a decade with the decade average as base. Application of these relatives to our accepted decade aggregates would yield estimates of annual housing production coordinated with both decade level and annual patterns of movement. It would be unwise to apply the indexes implicit in any one set of annual estimates on record to trace out the course of residential building over the century. None of them could be totally disregarded; none could be wholly trusted. Each of the estimates attempted represents a tendency in building which should be reflected in the totals and indexes. Hence it seemed desirable to work decade by decade and to assemble sets of decade indexes corresponding to different tendencies in residential building—either by area or city size class—or with biases that would be offsetting. In assembling the component indexes and determining their respective weights, it was necessary to keep in mind decade fitting at decade endings. The decade indexes and the final estimates are shown in Table 15.

For the decade of the 1930's, the index pattern of the BLS-NBER series, the outcome of skilled manipulation of a broad permit reporting system, was applied to our estimated decade total to yield annual estimates. The only alternative pattern by which to gauge its adequacy was that afforded by the 1940 Census vintage report which, for the thirties, spelled out for each year the reported production of dwelling units allegedly built in that year and surviving to the 1940 Census date. Since in census-reported age distribution the reported ages bias toward round figures, the reported 1930 production is excluded from our decade index comparison. We show in Chart 13 the index patterns for 1931-39 vintage residential production by city-size class, along with the official series. The fit is close. Hence we accepted the pattern of the official indexes and

Estimation of Decade Annual Indexes

CHART 13

Decade Indexes of Dwelling Units, 1931-39

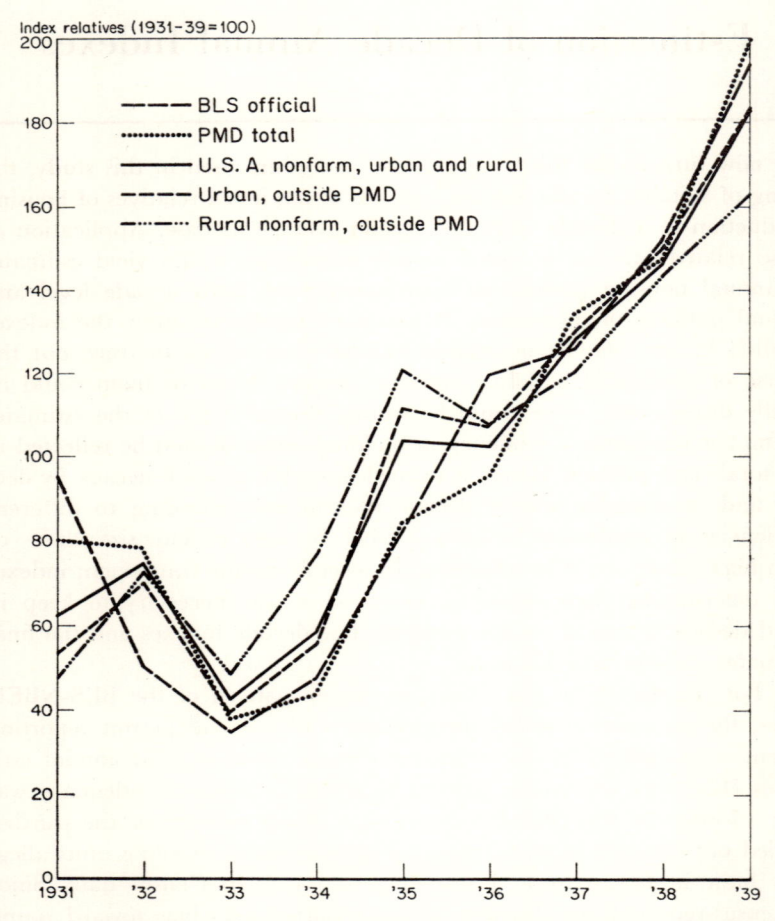

Source: BLS figures from *Construction During Five Decades*, 1953, p. 3; the other four series obtained from 16th Census, *Housing, Characteristics by Type of Structure*, 1945, pp. 3-7.

Estimation of Decade Annual Indexes

TABLE 15

ESTIMATED PRODUCTION OF RESIDENTIAL HOUSEKEEPING UNITS, 1840-1939

Year	Decade Indexes	Housing Production (000)	Year	Decade Indexes	Original Housing Series Production (000)	(000)
1840	69.01	36	1890		328	278
1841	79.02	41	1891		298	252
1842	65.45	34	1892		381	323
1843	65.72	34	1893		267	226
1844	73.32	38	1894		265	225
1845	87.98	46	1895		309	262
1846	111.16	58	1896		257	218
1847	137.68	72	1897		292	247
1848	134.06	70	1898		262	222
1849	157.71	82	1899		282	239
1850	78.28	94	1900	54.75		230
1851	90.28	109	1901	78.58		330
1852	100.70	121	1902	76.09		320
1853	110.53	133	1903	82.65		347
1854	114.72	138	1904	96.00		403
1855	115.20	138	1905	127.42		535
1856	112.13	135	1906	124.65		524
1857	118.08	142	1907	112.26		471
1858	86.09	103	1908	110.04		462
1859	81.24	98	1909	137.22		576
1860	87.80	93	1910	113.13		477
1861	71.32	76	1911	111.87		472
1862	74.04	79	1912	120.93		510
1863	69.35	74	1913	113.10		477
1864	56.19	60	1914	110.73		467
1865	76.62	81	1915	112.40		474
1866	107.65	114	1916	110.58		467
1867	131.86	140	1917	68.97		291
1868	156.70	166	1918	40.14		169
1869	159.06	169	1919	98.38		415
1870	118.63	158	1920		217	250
1871	140.54	188	1921		491	491
1872	125.76	168	1922		767	767
1873	130.39	174	1923		950	950
1874	92.69	124	1924		963	963
1875	95.53	127	1925		1,048	1,048
1876	76.48	102	1926		1,010	1,010
1877	77.11	103	1927		837	837
1878	60.63	81	1928		741	741
1879	67.16	90	1929		473	573

(continued)

Estimation of Decade Annual Indexes

TABLE 15 (concluded)

Year	Decade Indexes	Housing Production (000)	Year	Decade Indexes	Original Housing Series Production (000) (000)
1880	48.76	127	1930		330 485
1881	73.61	191	1931		254 373
1882	80.19	208	1932		134 197
1883	91.43	237	1933		93 137
1884	94.89	246	1934		126 185
1885	102.73	267	1935		221 325
1886	107.00	278	1936		319 469
1887	125.61	326	1937		336 494
1888	123.31	320	1938		406 597
1889	141.61	368	1939		515 757

NOTES

1840–59: Decade totals of 520,000 and 1,202,000 units distributed on the basis of decade indexes made up of the geometric average of Riggleman (Isard-adjusted) and Ohio building value. Ohio figures from unpublished tax increments (see discussion pp. 75 ff.); Colean and Newcomb, *Stabilizing Construction,* p. 227.

1860–89: Decade totals of 1,061,000, 1,333,000, and 2,597,000 units for the sixties, seventies, and eighties, respectively, were distributed by the geometric average of the following indexes equally weighted: (1) Ohio statewide, (2) Riggleman-Isard adjusted, deflated by a Riggleman cost of building index, and (3) Long index-number residential building permits (Long, *Building Cycles,* pp. 227-228; Colean and Newcomb, *Stabilizing Construction,* p. 226; Riggleman, *Variations in Building Activity,* pp. 257-258; Table 16 following).

1890–99: Blank's yearly estimates reduced by 15.3 per cent. Yearly totals in Grebler, Blank, and Winnick, *Capital Formation,* p. 332, Table 15.

1900–19: Chawner's decade estimates of 4,200,000 and 4,220,000 distributed by index relatives (1900-19=100) made up of simple average of Blank's and Chawner's series (Blank, *Volume of Residential Construction,* p. 27, Table 13.

1920–29: Blank's new series at 7,497,000 for the decade (compared with official series at 7,034,000). Figures for 1920 adjusted by 33,000; figures for 1929 raised by 100,000. The adjustment for 1929 was for "linkage" only and was for that reason arbitrary. The need for adjustment indicates a hiatus in the sets of decade totals and yearly indexes. *Ibid.,* p. 59.

1930–39: Official BLS-NBER yearly estimates (*Construction During Five Decades,* 1953, p. 3), raised by 47 per cent to make a 4,019,000 decade total.

Sum of indexes by decades differs from 1,000 because of use of geometric averages and because of rounding. For the same reasons, production totals by decade also differ slightly from the decade aggregates that were distributed.

Estimation of Decade Annual Indexes

raised them by a decadewide adjustment to offset the underestimation previously noted.

With Blank's decade total for the twenties, the decade index that went along with it was accepted.[55] As Blank noted, this index corresponds closely with that of the official series. I departed from his index for 1929 to smooth the transition, and for 1920 because his estimate for that year in the light of available information seemed unduly low.[56]

For the decades of the 1900's and 1910's, Chawner's decade totals were used, as previously noted, but not the unmodified index patterns that went along with the totals. Blank's decade indexes involved decade totals which, like Chawner's, were about the same in magnitude for the two decades though resting at a lower level. Blank's decade indexes reflect a more intensive working of a wider mass of empirical materials. But some of the pattern implicit in Blank's indexes must have been generated by the underestimation of level running through them. Under the circumstances it seemed reasonable to average the two sets of decade indexes. The average of the resulting pair of indexes is shown in Table 15.[57]

For the 1890's, Blank's index was used, since it was drawn from a city coverage much larger than other available indexes were. By 1890, his index is based upon 25 cities accounting for 14.5 per cent of total nonfarm population. In contrast, Long's residential permit index for 1890 covered only 12 cities. The Riggleman index for 1890 has the requisite coverage (32 cities) but it measures nonresidential as well as residential building and it is cast in value terms. Only the Ohio residential building index was suited for averaging with the Blank index. But since the Blank index was adjusted for nonurban building and since the two idexes corresponded closely (see Chart 14), the Blank index was accepted for the nineties.[58]

---

[55] For details see Blank, *The Volume of Residential Construction*, p. 59.

[56] Estimates for the year 1920 compare as follows: Wickens (1941), 247; Chawner (1939), 300; Blank (1954), 217; BLS (1959), 247. (Blank, *Volume of Residential Construction*, Table 13, p. 59; Wickens, *Residential Real Estate*, p. 49; Chawner, *Residential Building*, p. 13; *Construction During Five Decades*, p. 3.)

[57] See the detailed discussion of the Chawner and Blank indexes in Blank, *Volume of Residential Construction*, pp. 25-28.

[58] With the index relative, 1890-99 = 100, the combined decade indexes are as follows for the years beginning 1890:

	1890	1891	1892	1893	1894
Blank, alone	115.6	101.4	129.6	90.8	90.1
Blank-Ohio average	102.7	109.6	127.6	100.7	89.3

	1895	1896	1897	1898	1899
Blank, alone	105.1	87.4	99.3	89.1	95.9
Blank-Ohio average	102.7	89.0	93.5	87.6	99.4

Estimation of Decade Annual Indexes

CHART 14
Decade Indexes for the 1890's

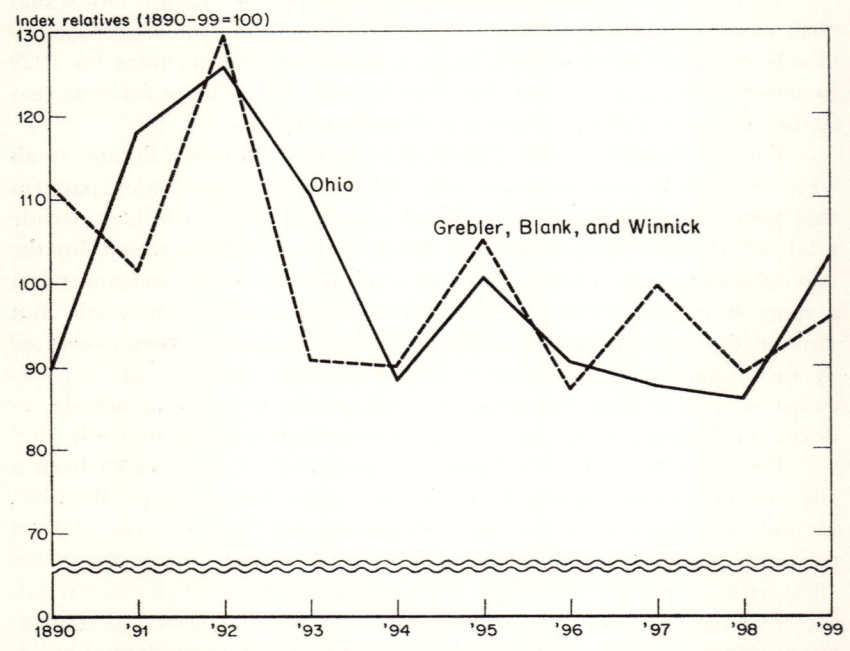

Index relatives (1890-99=100)

[Chart showing two lines—Ohio and Grebler, Blank, and Winnick—plotted from 1890 to 1899, with values ranging approximately between 85 and 130.]

Source: Grebler, Blank, and Winnick, *Capital Formation*, p. 332; unpublished Ohio yearly totals.

For the three decades between 1860 and 1890, difficulty was experienced in devising appropriate annual indexes. None of the standard available measures had a claim to special validity in those three decades. Blank's index was too limited in coverage to be used alone for the 1880's. The Ohio experience is likewise unsuitable as a sole measure of nationwide patterns of movement, since Ohio overresponded to the rise of the sixties and underresponded to the boom of the eighties. The annual statewide Ohio returns for residential production also include farm dwellings, which played a variable role over those three decades. Newly erected farm dwellings are estimated to account for 27, 50, and 13 per cent of aggregate decade production of houses during the 1860's, 1870's, and 1880's, respectively (see Table 11, columns 1 and 6). There is no reliable way to remove the influence of farm construction from annual aggregate statewide returns. Aggregation of returns separately for highly urbanized

Estimation of Decade Annual Indexes

and other counties showed that the level of production in the highly urban counties receded more during the seventies—relative to the late sixties and eighties—than it did in other counties (see Table 16). The highly urban counties reached a peak in 1868 which nearly matched the 1873 peak; for other counties the boom in the early seventies carried

TABLE 16

NUMBER OF URBAN AND OTHER DWELLING UNITS PRODUCED IN OHIO, ANNUALLY, 1860-89
(thousands)

Year	Urban	Others	Total
1860	1,864	5,821	7,685
1861	1,390	4,948	6,338
1862	1,306	3,652	4,958
1863	1,405	3,933	5,388
1864	1,631	3,537	5,168
1865	2,832	5,853	8,685
1866	4,006	8,253	12,259
1867	4,004	9,415	13,419
1868	4,233	9,997	14,230
1869	3,257	8,103	11,360
1870	2,775	7,164	9,939
1871	3,467	9,134	12,601
1872	3,658	11,514	15,172
1873	4,413	12,580	16,993
1874	4,318	10,657	14,975
1875	3,839	11,538	15,377
1876	2,574	8,574	11,148
1877	2,674	8,331	11,005
1878	1,874	6,534	8,408
1879	2,586	8,161	10,747
1880	2,523	7,302	9,825
1881	4,567	14,848	19,415
1882	4,962	14,392	19,354
1883	5,901	14,834	20,735
1884	6,194	12,373	18,567
1885	5,840	10,156	15,996
1886	5,033	9,916	14,949
1887	8,342	12,423	20,765
1888	8,897	11,862	20,759
1889	8,259	12,766	21,025

NOTE: Urban totals for five counties containing large cities: Hamilton (Cincinnati), Franklin (Columbus), Montgomery (Dayton), Cuyahoga (Cleveland), Lucas (Toledo); the "others" make up the rest of the state with its 83 counties. The total is taken from NBER series 0147.

Estimation of Decade Annual Indexes

residential building to levels nearly a third above the peak of 1868. On the other hand, the long-swing contraction was equally marked in both categories; and total specific amplitude of residential building tends to vary, if at all, inversely with degree of urbanization.[59] Thus while our total statewide building index is serviceable, possible merit may be found in separate use of its highly urbanized and other components.

The Long index before 1890 reflects the building experience of a few metropolitan centers. It is excessively weighted with New York (Manhattan) experience and fails to allow adequately for secular growth. The Riggleman index as adjusted by Isard is also metropolitan in character but during those decades it is much more broadly based and less dependent upon eastern seaboard building patterns. The index measures the value of total urban building, not of residential units; and small towns and cities are not adequately weighted in the index. Besides these major indexes, we have available separate city returns for Manhattan from 1868 and for Chicago from 1860 on.

Two possible approaches were explored in detail. First, an effort was made to assess index availabilities for each decade and to select an appropriate set of indexes with changing weights. The judgment involved in selecting components and assigning weights was inherently opportunistic and marked by some arbitrariness, but variation of weights and components by decades improved judgment for each decade. The three sets of decade indexes developed in this fashion are shown in the next tabulation.[60]

[59] Below are urbanization indexes, chronologies, and specific amplitudes for statewide Ohio and its sample groups.

Sample Group	Peak	Trough	Amplitude Specific Cycle Fall	Urban Population as Per Cent of Total, 1920	Farm Mortgages as Per Cent of Total Value, 1885
I	1874	1878	85.2	93	12
II	1868	1878	86.7	83	39
III	1868	1878	83.4	77	37
IV	1872	1878	102.5	43	57
V	1875	1878	118.7	20	84
State	1873	1878	74.5	64	49

[60] For the decade of the eighties Blank's index was combined with the Ohio statewide index, with Long's index of residential permits, and with Riggleman's per capita building permits (see Chart 15 and Table 17). An alternative average of a set of indexes including Chicago and Manhattan corresponded in pattern with the simple average of the four indexes. The high amplitudes of the Long index should offset the limited downward bias in the population adjustment of the Riggleman index. For the decade of the seventies, a broadly based Blank index was no longer available, and greater reliance had to be placed on the Riggleman index with its broad coverage. It was assigned a double weighting. The Riggleman deflated index was used, since in the decade a profound price

Estimation of Decade Annual Indexes

Component Activities	Weight Allowed in Decade Index		
	1860's	1870's	1880's
Ohio, statewide			1
Riggleman, per capita			1
Riggleman, per capita deflated	1	2	
Blank, residential units			1
Long, number residential permits	1	1	1
Ohio, urban	1	1	
Ohio, other	1	1	

Tables 17-19 and Charts 15-17 show the behavior of the several indexes and their averages. Application of the averages to our decade totals yielded a set of annual building totals, reproduced in Chart 18, "shifting weight variant." The second procedure used to derive decade indexes for 1860-89 involved a constant set of series with unchanging weights for the three decades. This eliminated the arbitrariness involved in shifting indexes and weights by decades though it entails a less efficient use of available information. The most serviceable index measures running for the three decades were found to be Ohio, statewide; Long, number of residential permits; and Riggleman-Isard, deflated value of building permits. It was necessary to deflate the Riggleman-Isard series since in the period covered a profound price revolution distorted value calculations cast in current prices. The three series reflect, in different ways and with different degrees of emphasis, divergent tendencies of United States building experience. Fortunately, the main outlines of movement being shared, the pattern of weighting was not too difficult. In the absence of measured criteria, each index was weighted equally and averaged geometrically. As Chart 18 shows, the constant weight and variable weight series conform closely, facilitating selection of the constant weight series because of the simplicity of its design.

---

revolution distorted value calculations cast in current prices. It was felt that the upward bias in a metropolitan building series and inclusion of Long's high-amplitude and limited-coverage index offset the downward bias inherent in the per capita adjustment. Since the Riggleman index in the seventies did not cover Ohio cities, separate averaging of the Ohio urban index, with its five large urban counties, was deemed justifiable. The Ohio nonurban index was included to represent smaller urban patterns of movement. A test computation of another plausible pair—Riggleman and Ohio nonurban—displayed a parallel pattern. The broader based average was selected (for indexes, see Table 18 and Chart 16). For the decade of the sixties, the combination of Riggleman, Long, and Ohio urban and nonurban was used. Because of its reduced coverage of only six cities (four eastern seaboard, plus Milwaukee and Indianapolis) and its predominant reflection of eastern seaboard urban experience, the Riggleman index was weighted only once. While inclusion of two Ohio indexes may not be fully defensible, both reflect patterns of economic activity running through the whole central northwest territory. In the absence of better information, their inclusion seemed justifiable (see Table 19 and Chart 17).

Estimation of Decade Annual Indexes

TABLE 17

DECADE INDEXES OF RESIDENTIAL BUILDING FOR THE 1880's
(1880-89=100)

	1880	1881	1882	1883	1884	1885	1886	1887	1888	1889
1. Ohio, statewide	54.2	107.0	106.7	114.3	102.4	88.2	82.4	114.5	114.4	115.9
2. Riggleman, per capita	62.3	76.7	90.9	98.9	93.3	104.4	115.1	118.4	109.4	130.6
3. Blank	50.8	57.7	60.5	83.8	100.6	113.8	130.1	144.3	118.4	140.1
4. Long, residential	40.7	57.8	64.2	79.2	89.9	113.5	128.5	137.0	130.6	158.5
5. Geometric average of lines 1 to 4	51.4	72.3	78.3	93.1	96.4	104.4	112.2	128.0	117.9	135.4

SOURCE, by line

1. Table 16.
2. Riggleman, *Building Cycles*, pp. 258-60.
3. Blank, *Volume of Residential Construction*, p. 42, Table 11.
4. Long, *Stabilizing Construction*, p. 228, App. B, sect. 3.

TABLE 18

Decade Indexes of Residential Building for the 1870's

	1870	1871	1872	1873	1874	1875	1876	1877	1878	1879
1. Riggleman permits per capita, (1913 dollars)	138.9	147.9	115.3	111.8	95.6	96.4	79.9	73.9	64.8	75.5
2. Long, residential, number of permits (1870-79=100)	139.1	169.6	126.1	126.1	89.1	89.1	73.9	80.4	56.5	50.0
3. Ohio, number of urban residential (1870-79=100)	86.2	107.7	113.7	137.1	134.2	119.3	80.0	83.1	58.2	80.4
4. Ohio, number of nonurban residential (1870-79=100)	76.1	97.0	122.2	133.6	113.1	122.5	91.0	88.5	69.4	86.6
5. Geometric average, lines 1 to 4	111.9	131.1	118.4	123.6	104.3	103.9	80.8	79.8	62.6	72.4

Source: Same as for Table 17.

Estimation of Decade Annual Indexes

TABLE 19
DECADE INDEXES OF RESIDENTIAL BUILDING FOR THE 1860's

	1860	1861	1862	1863	1884	1865	1866	1867	1868	1869
1. Riggleman, permits per capita (1913 dollars)	126.0	83.2	72.4	87.1	52.1	74.9	101.7	118.0	136.6	147.8
2. Long, residential, number of permits (1860-69=100)	85.7	62.9	91.4	88.6	45.7	62.9	88.6	125.7	160.0	188.6
3. Ohio, number of urban residential (1860-69=100)	71.9	53.6	50.4	54.2	62.9	109.2	154.5	154.4	163.3	125.6
4. Ohio, rest of state, number of residential (1860-69=100)	91.7	77.9	57.5	61.9	55.7	92.2	129.9	148.2	157.4	127.6
5. Geometric average lines 1 to 4	91.9	68.4	66.2	71.3	53.7	83.0	116.0	135.7	154.0	145.4

SOURCE: Same as for Table 17.

Estimation of Decade Annual Indexes

CHART 15

Decade Indexes of Urban Residential Building for the 1880's

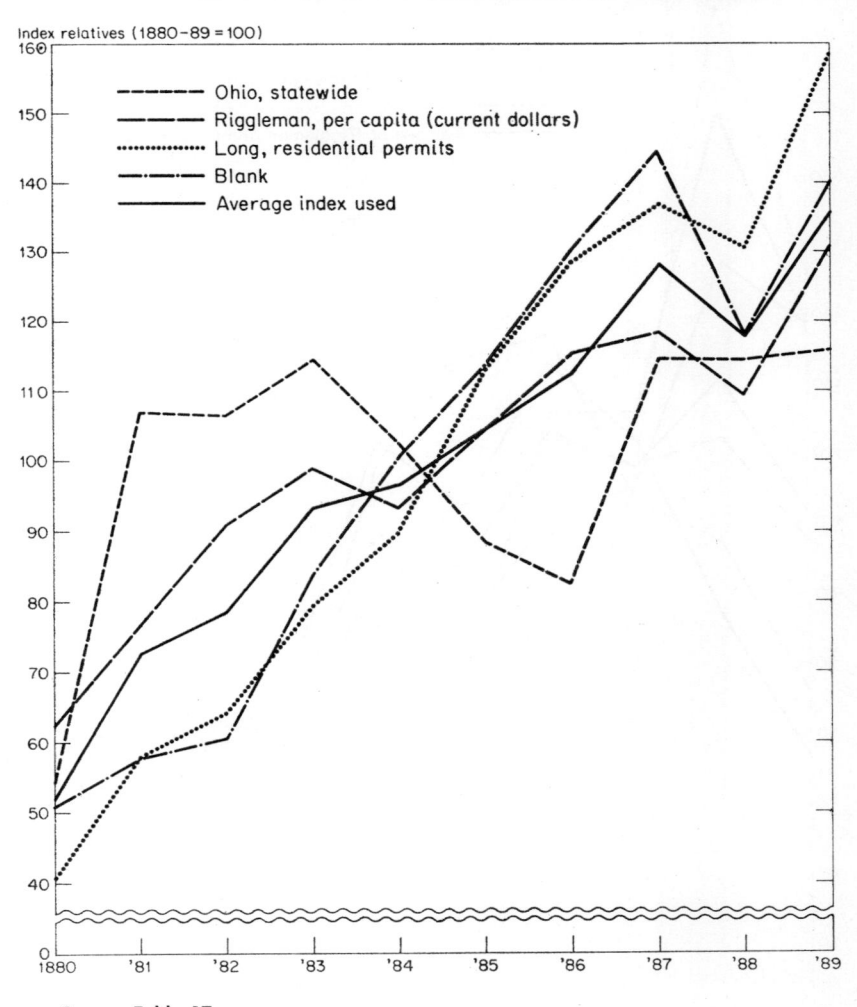

Index relatives (1880-89 = 100)

- Ohio, statewide
- Riggleman, per capita (current dollars)
- Long, residential permits
- Blank
- Average index used

Source: Table 17.

Estimation of Decade Annual Indexes

CHART 16

Decade Indexes of Urban Residential Building for the 1870's

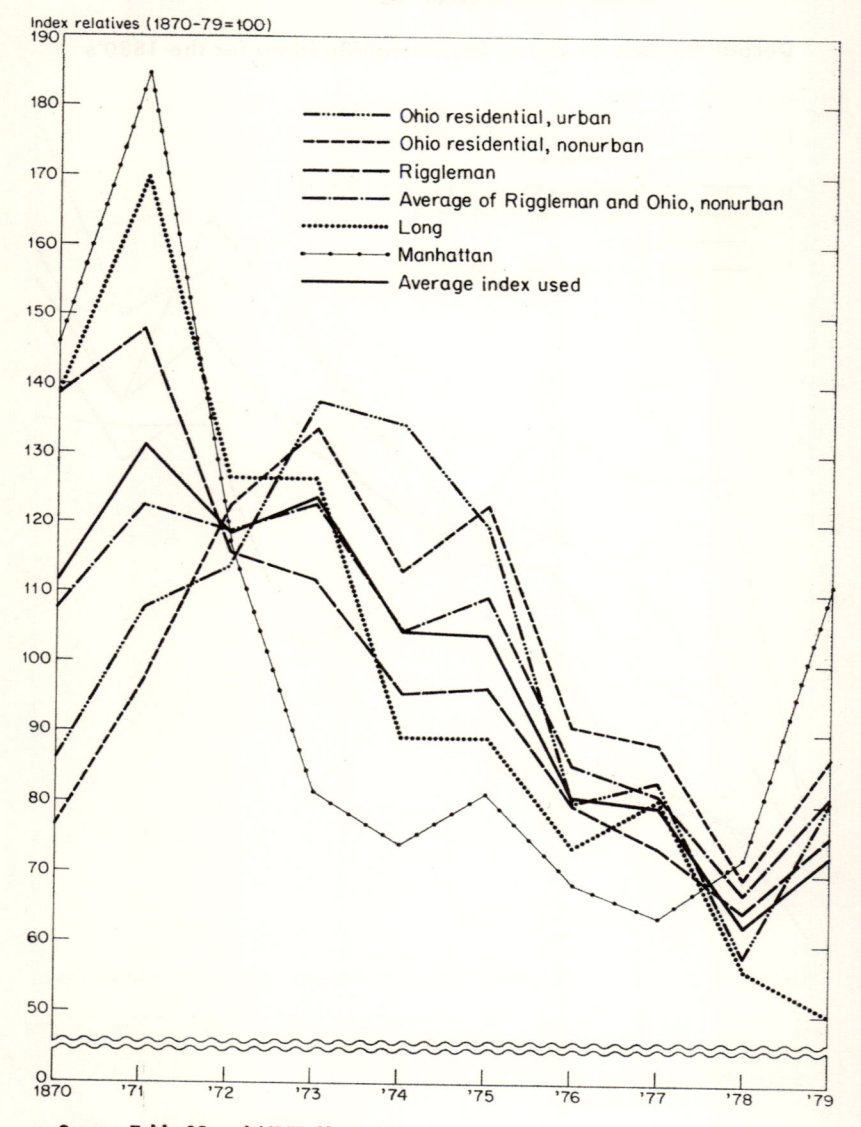

Source: Table 18 and NBER files.

Estimation of Decade Annual Indexes

CHART 17

Decade Indexes of Urban Residential Building for the 1860's

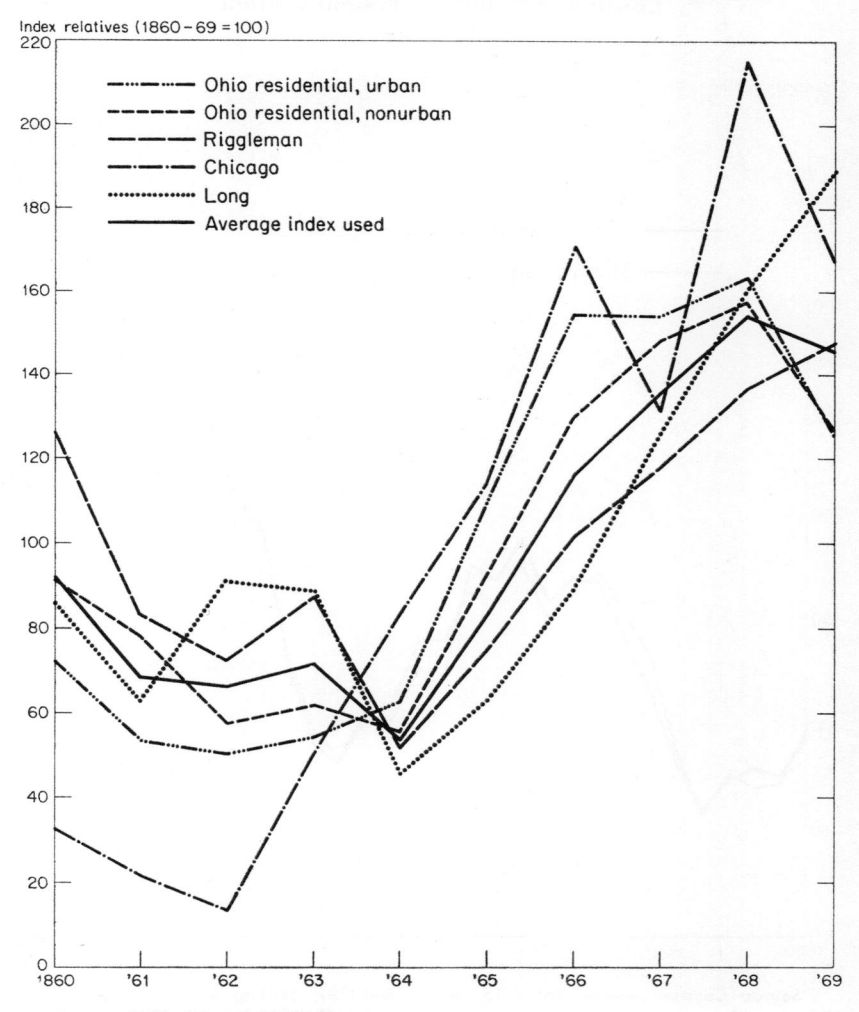

Index relatives (1860-69 = 100)

— Ohio residential, urban
------ Ohio residential, nonurban
— — Riggleman
—·— Chicago
········ Long
—— Average index used

Source: Table 19 and NBER files.

Estimation of Decade Annual Indexes

CHART 18

Estimated Annual Residential Production, Constant and Shifting Weight Variants, 1860-89

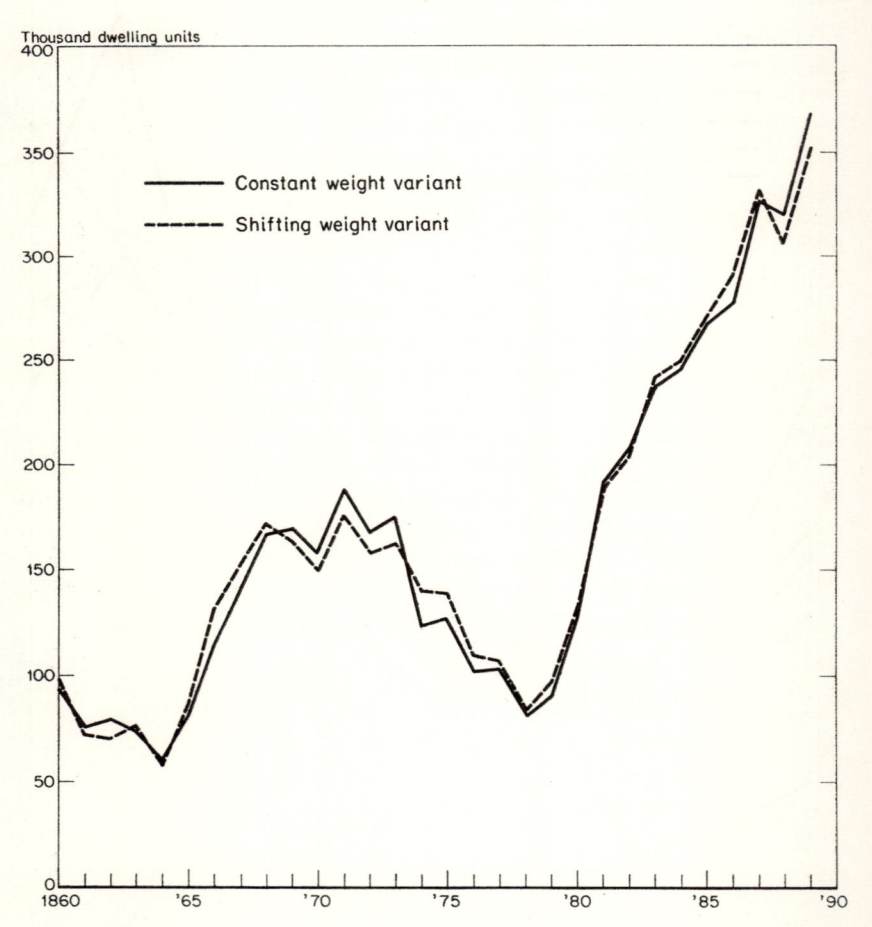

Source: Constant weight: Table 15, years 1860-1889. Shifting weight: Tables 17, 18 19, geometric average, multiplied against decade totals listed in notes to Table 15.

Estimation of Decade Annual Indexes

In applying annual indexes to the decade total for the 1860's allowance was made for replacement of Civil War losses in housing. Of the estimated 1860 stock of nonfarm dwellings, an estimated 22 per cent or 628,000 were in the South. Civil War damage was intense in certain areas: the upper valley of the Shenandoah; the country between Washington, D.C. and Richmond; cities such as Atlanta, Columbia, Nashville, Charleston; the belt of territory ravaged by Sherman's army; and other areas of concentrated fighting.[61] By a rough guess on the basis of estimation of the populations involved, housing production for 1866-69 was raised by 32,000 units (see Table 15 and notes).

We turn now to the estimation of the annual series for 1840-59. The only comprehensive annual indexes of building available are those derived from Ohio and the Riggleman study. The Riggleman index for those decades reflects building rates in the eastern seaboard urban communities of Manhattan, Boston, and Washington. (Three other cities were added to the index, but either late in the surveyed years or with little effect on the totals.) The Ohio index for the two decades was derived from the 1840-60 record of increments of real property assessments in Ohio municipalities converted into building value with the aid of regression analysis of total value of new building and annual increments of the assessed value of municipal (town, village, and city) real property between 1858 and 1889. The regression involved problems of two separate classes. There was, first, the need to adjust the correlated variables for comparability and, second, to design an adequate regression relationship which "fitted" the observed relationship.

The total value of building as initially reported in Ohio building statistics, including farm and nontaxable public building, was utilized as

[61] Sherman's march through Georgia and the Carolinas allegedly was accompanied by much burning and destruction, "not a building [was left] on the railway from Macon to Savannah" (O. Eisenschiml and R. Newman, *The Civil War*, New York, Grosset & Dunlap, Vol. I, 1956, p. 658; E. M. Coulter, *The South During Reconstruction 1865-1877*, Baton Rouge, Louisiana State University Press, 1947, p. 1). The upper valley of the Shenandoah, according to J. B. McMasters, "was a scene of desolation. . . . Beyond Harrisonburg, toward Staunton, was the region laid waste by Sheridan, and there, within a circle five miles in diameter, scarcely a house, barn, mill or building of any sort was standing." The country between Richmond and Washington was "like a desert." Homes of Union volunteers in Tennessee and northern Georgia "had been ruined by rebel armies or guerillas." Charleston was allegedly a city of "ruins" (*A History of the People of the United States During Lincoln's Administration*, New York, D. Appleton, 1927, pp. 637-640). Columbia, S. C., was burned under official orders. A tentative number of 32,000 dwelling units was allowed for war loss. According to general historical accounts rebuilding was speedy (see Coulter, *op. cit.*, p. 255 ff.). Hence the war replacement was spread over 1866-69 and moderately peaked in 1866.

one of the correlated variables. Property-tax assessment reports refer to nonfarm real property, excluding public or quasi-public property. Despite these elements of noncomparability, the levels of building and tax-assessment increments, as shown in Chart 19, were closely aligned throughout the entire period. The nonfarm taxable building predominated in total building values, even in the early period of the comparison; and possibly the relative movement of farm and public building was mutually offsetting. At a later stage of the research, another regression was prepared between assessment increments and building values, fully adjusted for comparability. The assessment increments were for all real property (farm and nonfarm) and the taxable building excluded all public and quasi-public building (which would not be represented in the assessments). The fit during the "regression" period is somewhat closer in the second case.[62] These more closely matched variables apply to all building and property increments and, in the early years 1837-55, the changes in farm real estate assessment dominate the assessment totals. The relative share of nonfarm assessments was rising rapidly during those years. Since the present quest is for a nonfarm building index, it seemed desirable to utilize the regression which applied only to assessment increments in municipal property.

The assessment increments measure the net resultant of the assessed value of (1) old buildings dropped from the tax rolls because of demolition or destruction, (2) assessed values of newly erected structures or improvements added to the tax rolls, and (3) revaluations of properties already on the rolls. Revaluations disturb the relationship between net new building and net change in assessments. Before 1846, the disturbances were marked owing to laxity in assessment and weak central supervision. After 1846, disturbances due to revaluations were scaled down, since revaluations were concentrated in scheduled reappraisal years set at first for 1853 and 1859, thereafter, decennially. The upward shift of values resulting from revaluations was sizable in 1853 and in 1870. Assessed values for identical or nearly identical properties were raised 61 per cent and 60.9 per cent, respectively, reflecting the inflations of 1846-53 and 1861-66. The revaluation of 1859 showed a 5.5 per cent fall in the level of real estate values fixed in 1853. After 1870, the standard of assessment seemed fixed and resulted in a widening gap between assessed and market values. Outside the reappraisal years, revaluations were permitted only

[62] See charts of these matched variables and description of regression in my paper, "Value of Nonfarm Building, Residential and Nonresidential, U.S.A., Annually, 1850-1939," submitted to the Conference on Research in Income and Wealth, Chapel Hill, North Carolina, September 4-5, 1963.

Estimation of Decade Annual Indexes

CHART 19
Ohio Building, 1840-89

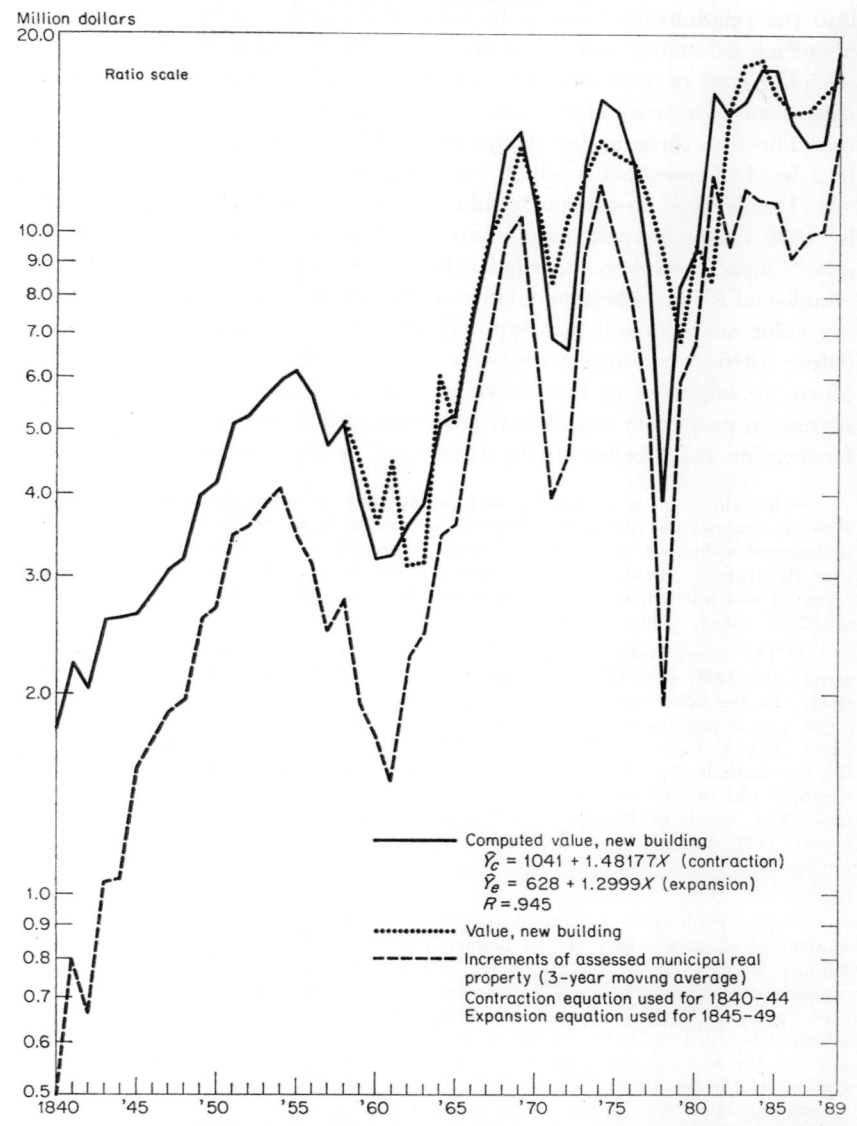

- Computed value, new building
  $\hat{Y}_c = 1041 + 1.48177X$ (contraction)
  $\hat{Y}_e = 628 + 1.2999X$ (expansion)
  $R = .945$
- .... Value, new building
- ------ Increments of assessed municipal real property (3-year moving average)

Contraction equation used for 1840-44
Expansion equation used for 1845-49

Source: Assessments from state auditor's annual reports, as reported by letter from auditors' offices, and as compiled for this study by the Secretary of the State of Ohio. Value of new building was extracted from annual reports of the Secretary of State. These data are unadjusted for assessment undervaluation.

Estimation of Decade Annual Indexes

in exceptional circumstances and appear to have affected only small fractions of assessed property on the rolls.[63] Disturbances thus introduced into the relationship between building and assessment increments could be offset or neutralized by smoothing with a 3-year moving average.

The level of assessment or relationship of assessment values to market values was, for years after 1846, accepted as originally prepared for tax use. The first three major reappraisals—1846, 1853, and 1859—resulted in a level of assessments which closely approximated market value.[64]

The level of assessment gradually receded after the Civil War and by 1890 hardly covered more than two-thirds of the value of the property.[65] Since, however, recorded values for new building reflected values established for tax assessment, correction of both tax-assessment and building value statistics was not required. On the contrary, for the few years before 1846, when properties were evaluated for tax purposes at one-fourth to one-third of market value,[66] use of those values as a base for regression projection would have understated building in the early forties. Hence, for years before 1846, assessment values were doubled.[67]

---

[63] Revaluations were carried out periodically at intervals of 6 to 10 years. Pending reappraisal, old properties were assessed at fixed values. Hence, increments in assessed value, as noted in an official report, were "principally occasioned by new structures" (*Annual Report*, State Auditor, Ohio, 1865, p. 22). In the 1866 report it was noted that "real estate stands taxed on the official value fixed upon it in 1859" (*ibid.*, 1866, p. 20).

[64] The census estimates of true value of all taxable property (real and personal) for 1850 and 1860 were understated by 19.6 and 14.0 per cent, respectively. In the contemporary view of the Ohio statistics commissioner, understatement was concentrated in personal property (see Ohio Statistics Commissioner, *Sixth Annual Report*, 1862, p. 24). His finding was confirmed by deed surveys for the periods Apr.–Oct. 1853 and the year ending 1859, showing that assessment of sold properties ran to 87.4 and 101.0 per cent of market value (*Proceedings 1853*, Board of Equalization, Columbus, Ohio, 1854; *Ohio Executive Documents*, 1859, part I, pp. 857-860).

[65] See data and reports collected in my paper, "Value of Nonfarm Building."

[66] One responsible judgment was that, before 1846, assessed values ran to a quarter of market values (E. L. Bogart, *Financial History of Ohio*, University of Illinois Studies, 1912, p. 210). The 1846 reassessment boosted the values of assessed urban property by 256 per cent. The Ohio statistics commissioner noted that "prior to 1846 the assessments fall so much below the real values that they afforded but little criterion of the actual wealth" (*loc. cit.*). On the 1846 revamping, see the history and appraisal in Bogart, *Financial History of Ohio;* and *The Passing of the Frontier, 1825-1850*, C. Wittke, ed., Columbus, Ohio State Archeological and Historical Society, 1941, pp. 425-430.

[67] Actual annual assessment increments for real property located within municipal corporations, the increments adjusted for undervaluation before 1846,

Estimation of Decade Annual Indexes

The form of the regression presented difficulties. A trial estimation was made by fitting a linear regression line between new-building and tax-assessment increments. The results showed a systematic bias in the estimated value; on the expansion phase, there was a tendency toward overestimation and, on the contraction phase, underestimation reflecting a tendency for assessment revaluation of old properties to exaggerate swings. An attempt to allow for this behavior by using a multiple-linear regression function, including annual percentage change as a variable, did not produce workable results. The coefficient of correlation was .69. After tests of several other multiple-regression linear functions, also with unsatisfactory results, the correlation formula finally employed was an average of two simple linear functions, with and without constant terms, fitted separately to expansion and contraction phases. The coefficient of correlation of the resultant composite function is .94. The estimated value for new building from 1840 to 1857, along with actual values from 1858 on and smoothed assessment increments from 1840 on, are presented in Chart 19. Since the bases for regression are assessment increments in municipal real property, our projected values for new building reflect primarily nonfarm activity. The previously mentioned regression, based on increments of all real property including farms, involves a later trough for nonfarm activity in 1844 as against 1840, and a much dampened rise during the forties.

As indicated, neither the Riggleman nor the Ohio indexes are on as strong a footing for the decades before 1860 as for the decades following. In addition, they pertain to the total value of metropolitan and urban

and our three-year moving averages appear below. Since 1841 was a reappraisal year, the high negative value recorded for that year and the sizable gain for the subsequent year reflect the vagaries of shifting appraisal policy. For the calculations, the year 1841 was treated as zero.

	Actual Annual	Adjusted Annual	3-Year Moving Average
		(dollars in thousands)	
1839	441.3	882.6	—
1840	308.0	616.0	499
1841	−510.3	0	791
1842	878.3	1,756.6	667
1843	122.4	244.8	1,041
1844	561.5	1,123.0	1,051
1845	1,788.2	1,788.2	1,537
1846	n.a.	1,700.0	1,703
1847	1,620.1	1,620.1	1,873
1848	2,298.3	2,298.3	1,958

Estimation of Decade Annual Indexes

CHART 20

Decade Indexes for 1840-59

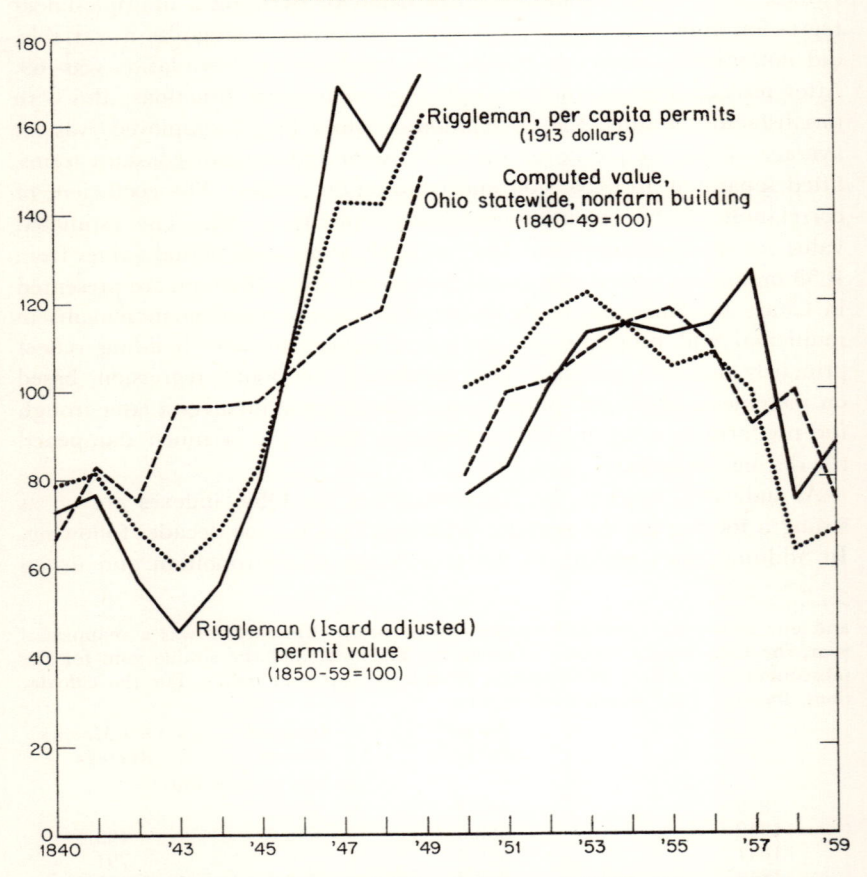

Riggleman, per capita permits (1913 dollars)

Computed value, Ohio statewide, nonfarm building (1840-49=100)

Riggleman (Isard adjusted) permit value (1850-59=100)

Source: Riggleman, *Variations in Building Activity*, pp. 287-288; Colean and Newcomb, *Stabilizing Construction*, p. 226, App. N, Table 2; estimates graphed in Chart 19.

Estimation of Decade Annual Indexes

building, which shares the swings of residential building, but often with variations in timing and amplitude. The amplitude and upward drift of the Riggleman index, as adjusted by Isard and used here, fit more closely to our accepted decade estimates of nonfarm residential building. Neither the Ohio nor the Riggleman series is deflated, partly because value shiftings in the period were difficult to measure accurately and partly because of doubt that Ohio assessments were adjusted to shifts in current costs. The resulting average of the two sets of decade indexes—Ohio nonfarm and Riggleman (Isard adjusted)—can lay out only the general course of residential building within those decades.

The two sets of decade indexes, along with Riggleman per capita permits in 1913 dollars, are shown in Chart 20. Since the movement is a common one, a geometric average (detailed in Table 15) of the two selected indexes was computed for use in projecting nationwide nonfarm residential building. An interesting check on the order of magnitude of our estimates for 1840-60 is afforded by the ratio of the 1857-59 recorded Ohio count of dwelling production to estimated nationwide dwellings. The Ohio share for 1857-59 is 6.3 per cent compared with 8.0 per cent for the decade of the sixties. The acceleration in Ohio building relative to nationwide levels was also found to characterize urban growth rates in Ohio and the nation (Chart 8). Another test for order of magnitude shows that residential building in 1844 in four large cities, accounting for 16.8 per cent of total estimated 1840 nonfarm population, amounted to 18.3 per cent of our nationwide nonfarm estimate for that year.[68] Our order of magnitude holds at least for this segment of the period.

---

[68] The editor of the *Cincinnati Miscellany* (Vol. II, 1864, p. 58) compiled the number of houses built in 1844 in Philadelphia, Boston, New York City, and Cincinnati (6,422). This is 18.3 per cent of our nationwide estimate for that year. The four cities had a census population in 1840 of 939,597. We estimate nonfarm population by scaling down our 1860 estimate (see footnote 39) in the ratio of the movement of nonagricultural labor force between 1840-60. Thus derived, the estimated nonfarm population in 1840 was 5.61 million of which our four-city share was 16.75 per cent.

6
Evaluation of Total Series

Despite their patchwork makeup by a combination of separately fashioned decade totals, the ten resulting sets of annual estimates of residential building fit together with little need for smoothing. Only one decade ending, that of 1929, was altered to smooth the connection. The resulting series is set forth in Table 15 and shown in Chart 21, together with the Long, Riggleman (Isard adjusted), and the Blank series.

All the major movements which run through these indexes, including short cyclical play, are reflected in ours. The new index is believed to catch more accurately the true secular drift and the relative play of long swings. The following tabulation of average long-swing amplitude and secular growth rate shows that the new index, compared with other measures, has low amplitude but relatively high secular growth. Five features of the new series call for detailed comment.

Series	Years Covered	Number of Long Cycles	Total Average Specific Amplitude	Long-term Rate of Growth (per cent per annum)
Riggleman-Isard index building permits	1862-1933	4	230.0	4.03
Gottlieb	1862-1939	4	200.4	2.74
Long index number residential permits	1856-1936	4	278.6	1.59[a]
Blank new dwelling units started	1889-1959	4	249.1	1.90

[a] Growth rate for Long's index of the total number of permits. (M. Abramovitz, *"Evidences of Long Swings in Aggregate Construction Since the Civil War,"* a National Bureau study, in preparation, Tables 7, 16).

First, the series shows a double peak between 1868 and 1871 with a sizable fall between, regarded here as corresponding with the nature of the post-Civil War expansion. The expansion in Ohio, where its course may be clearly traced, ebbed and flowed in course. As Table 16 and

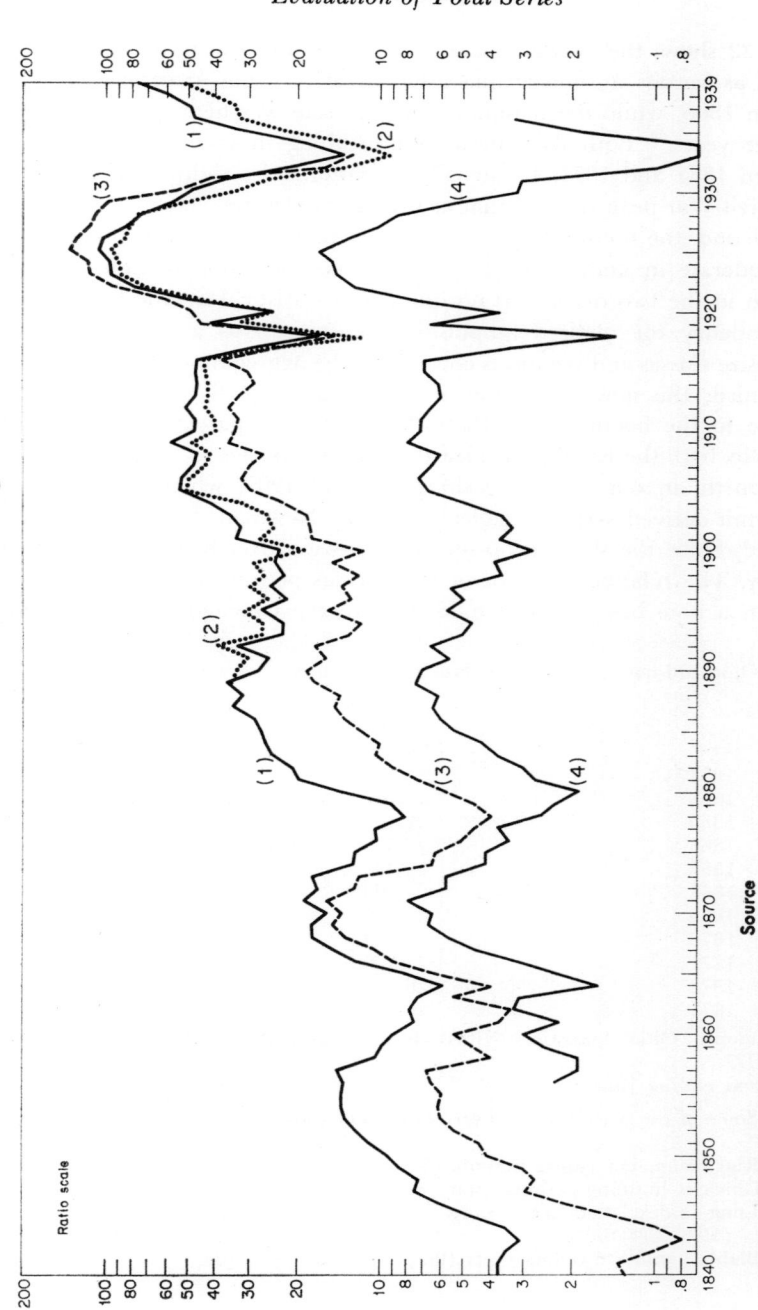

CHART 21
Estimates of United States Residential Building, 1840-1939

Evaluation of Total Series

Source
1. New series (10,000 units), Source Table 15.
2. Blank series (100,000 units), Blank, *Volume of Residential Construction.*
3. Riggleman-Isard index (1920-29 = 10,000), relatives in hundredths, source, Colean and Newcomb, *Stabilizing Construction,* p. 227.
4. Long (1920-30 = 100), relatives in tenths, source, *Historical Statistics,* 1949, p.173.

Evaluation of Total Series

Chart 22 show, the 1868-69 peak was sizable and the resulting ebb was almost as serious as the decline after 1873-74. Ohio marriages show a peak in 1866, while the national marriage rate was higher in 1867 than in later years.[69] Both data indicate an ebbing in household formation between 1867 and 1871. If our decade patterns are right, then the early post-Civil War peak rivalled that of the early seventies.

Second, the housing decline of the seventies and boom of the eighties are moderate in amplitude. The relative conformation of housing production in the two decades is no quirk of a statistical mosaic but reflects the tendency for realized amplitudes to dampen as a wider range of urban-size classes and regions is covered by the aggregate.

Third, the new index shows a substantial decline in the nineties relative to the boom of the 1880's and 1900's, a pattern of movement shared by both the Riggleman (Isard) and Long series.

Fourth, in our series the sharp drop of 1900, which characterizes all permit-derived series (Riggleman, Long, Newman, Blank), is greatly reduced. Since the sharp drop sets a long-swing trough it merits detailed enquiry. The relative behavior of the various permit-derived series is set forth in a note below.[70] The mild drop in the Chicago building permit

---

[69] Ohio and national marriage data are as follows:

Year	Ohio Number of Marriages[a]	Nation Marriages per 1,000
1858	21,352	
1863	19,187	
1865	22,404	
1866	30,579	
1867	29,230	9.6
1868	27,798	9.0
1869	25,625	8.9
1870	25,459	8.8
1871	24,849	8.8
1872	26,616	9.0
1873	26,692	9.0

SOURCE: Ohio Statistics, NBER files; *Historical Statistics*, 1949, p. 44, Series C77.

[a] Year ending June 30.

[70] Some of the permit-derived series for 1899-1901 follow:

	1899	1900	1901
Riggleman, per capita permits ($)	31.37	21.71	30.59
Chicago, building permits value ($ mill.)	20.9	19.1	34.9
Long, index, residence permits (1920-30=100)	38	28	36
Blank, number dwelling units (000's)	282	189	275

Evaluation of Total Series

CHART 22

Number of New Dwellings, Ohio, 1857-1914

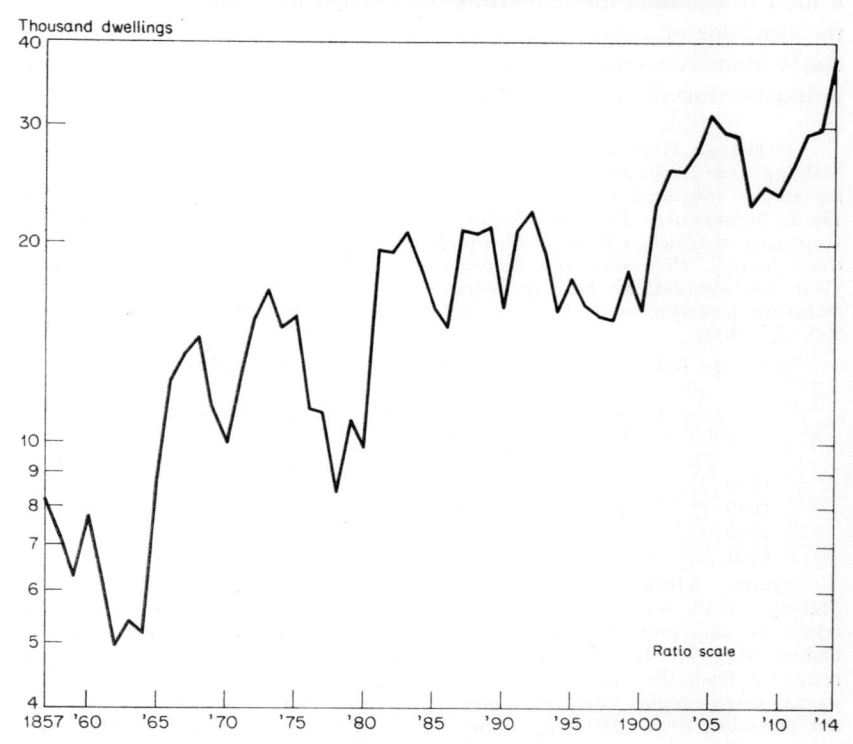

Source: NBER files of Ohio building data.

	1899	1900	1901
Newman, building index (1913=100)	70	46	66
Number of dwelling units			
Boston	3,700	2,378	2,569
Brooklyn	9,966	5,939	7,012
Manhattan	29,447	8,996	19,745
Detroit	1,278	1,323	1,620
San Antonio	222	323	629
St. Louis	1,390	1,287	1,850

SOURCE: M. Colean, *American Housing*, NY, 20th Century Fund, 1949, p. 408; GBW, p. 332; Riggleman, *Variations in Building*, pp. 260-1; Long, *Building Cycles*, App. B., p. 228; Newman, *Building Industry*, p. 63, Table IX, col. 4.

Evaluation of Total Series

series is inexplicable in view of the famous strike which paralyzed building in that city during most of 1900.[71] The report of building in Ohio shows a mild drop, while the nationwide production of building materials and the behavior of building prices testify, not to a drop in 1900, but to a rise.[72] Many cross-checking measures of building trade unemployment fail to indicate deterioration in 1900.[73] General building trade reports indicate

[71] Though Hoyt omits mention of the strike, it was well publicized in the building trade and made national news. A general contractor lockout in the building trades, instituted Feb. 5, 1900, paralyzed Chicago building until May 1901 (R. B. Montgomery, *Industrial Relations in the Chicago Building Trades,* Chicago, University of Chicago Press, 1927, pp. 28 ff.). For current notice see the building trade journal, *Carpentry and Building,* Jan. (editorial), Aug., Sept., and Nov. 1900. See also detailed hearings before the Industrial Commission (*Report of the Industrial Commission on the Chicago Labor Dispute of 1900,* Commission Report, Vol. 8, 1900).

[72] See the following record of behavior (values in dollars):

	Ohio, Value Total Building	Ohio, Number Dwellings	Shaw Deflated Construction Material	Building Material Prices (BLS index)
1898	24,327	15,237	1,341,569	39.6
1899	29,412	18,237	1,246,964	43.6
1900	29,748	15,967	1,425,045	46.2
1901	41,102	22,796	1,618,673	44,3

Source: NBER file 0147; unpublished adjusted building value; *Historical Statistics,* 1949, Series H50, p. 171, Series L22, p. 234. The 1900 Census editors' report on clay products manufacture noted that the census report for the year ending May 31, 1900, showed that the building trades were not yet "fully recovered from the panic of 1893" (1900 *Census of Manufactures,* p. 904). A special census report comparing manufacturing activity in the census year with the preceding year (1898-99) showed that brick and tile output in the census year was 16 per cent above that in 1898-99 (*ibid.,* p. lxi).

[73] See Paul Douglas, *Real Wages in the United States, 1890-1926,* Boston, Houghton Mifflin, 1930, pp. 135-6.

	Union Wage Rates, Building Trades (cents)	Average Workweek (hours)
1898	34.8	49.5
1899	36.1	48.9
1900	37.4	48.3
1901	39.1	47.5

In New York and Massachusetts, unemployment rates for 1900 showed a moderate percentage rise (though the peak was 1897): 1897, 32; 1898, 28.3; 1899, 20.9; 1900, 26.7; 1901, 17.8 (Douglas and A. Director, *The Problem of Unemployment,* New York, Macmillan, 1931). More conclusive on building trade unemployment in 1900 is a census count of unemployment for the year ending May 31, 1900. It showed, when compared with a similar count in 1890, higher percentage rates of unemployment during the same portion of the census year:

Evaluation of Total Series

marked slacks in Boston, Manhattan, and Chicago and a varied picture elsewhere, but no state of general decline.[74] Real estate indexes of market

	1899-1900	1889-90
Carpenters and joiners	41.4	31.8
Masons	55.5	42.9
Painters, etc.	42.4	31.1

As the census editor noted, the results are not comparable because of loose framing of the census questionnaire and failure to make allowance for normal seasonal unemployment or work interruptions due to weather, etc. Even in the more prosperous year that followed, 1901, a survey of annual earnings showed the wide distribution of full days worked by building trade workers (assuming a wage of 37.5 cents per hour, a fifty-hour week, and the average layoffs):

Average Estimate, Number Days Worked	Per Cent Full Days Worked 230 Bricklayers	1,095 Carpenters
27	0.4	0.5
40	0.4	0.9
67	1.3	2.5
93	7.4	6.3
120	8.3	12.1
147	8.7	18.2
173	19.1	23.5
200	25.6	22.4
227	20.9	7.1
253	11.7	4.7
293		1.9

(see *Cost of Living and Retail Prices of Food,* Department of Labor, p. 280 ff.). Reworking the 1889 and 1899 census results to allow for "normal" building trade unemployment, Douglas and Director (p. 30) found the percentages of unemployment were 8.8 per cent in 1889-90 and 12.8 per cent in 1899-1900, about what would be expected without a sharp drop in residential building in 1900.

[74] A "great and . . . general" falling off of building work was reported for New York City (*Carpentry and Building,* July 1900, p. 195), but it was ascribed to resistance to sharp 25 per cent increases in building material prices—characteristic of cyclical peaks. The monthly reports, "What Builders are Doing," for 1899 and 1900, reflect an uneven but not a dull picture for 1900.

Baltimore, "quiet though about normal," Aug.; Boston, off '99 by 30 per cent, July, but suburbs strong, Nov.; Cleveland, off '99 by 7 per cent; Columbus, Ohio, "very encouraging," up 25 per cent over '99, Nov.; Denver, "gratifying increase" over last year, Nov.; General Report, "it is well known that building operations have been on a comparatively small scale all over the country," Nov.; Hartford, Conn., "building quite active just at present," Nov.; Kansas City, Mo., "considerable amount of work in progress, . . . greater than at any time during the past ten years," Nov.; Los Angeles, "year expected to far exceed 1899," Aug.; Manhattan, off '99, 25 per cent, July; Memphis, Tenn., 27 per cent over 1899, Nov.; Milwaukee, off '99 by 35 per cent, July; New Haven, Conn., "showing more activity in building operations," July; New Orleans, "greater than ever before," Nov. and Dec.; Omaha, busier in July than in "a series of years," July; Pittsburgh, "new record," 33 per cent over '99, July; Portland, Maine, "building

Evaluation of Total Series

activity are positive or show only slight declines.[75] Retrospective surveys of general business conditions record in 1900 only a mild recession with no indication of a sizable fall in building.[76] From all this we conclude that the building permit indexes, with their overweighting of central city behavior, limited coverage, and inherent biases, erred badly in the 1900 count.

Fifth, in our series the pattern of movement around the ten-year period centering in 1900 (but exclusive of 1900 itself) shows a substantial rise in residential building in 1901-04 as compared with the rise in 1896-99. The ratios of total residential production in the later period to that in the earlier period, for our series and for related measures, are shown in the tabulation on the next page.

Thus, though our annual indexes were drawn entirely from BLS-NBER for the 1890's and substantially from that source for the 1900's,

---

operations are decidedly active," Nov.; Portland, Ore., "a large increase" over same months of previous years, Nov.; St. Louis, "very quiet," July; San Francisco, "good shape," "greater than for several years," little idleness, July and Nov.; Seattle, Wash., "now the most active building center of the Pacific Coast," Nov.; Tacoma, Wash., more than double 1899; Washington, Pa., "looking for something of a boom this fall," July; Worcester, Mass., "fair amount of building in progress," Nov.

[75] Thus, see the following:

		1899	1900	1901
(1)	Number mortgage recordings, 9 counties	49,333	46,271	46,161
(2)	Ohio, total mortgages (for year ending June 30)	78,191	79,026	81,861
(3)	Ohio total deeds (for year ending June 30)	111,879	111,877	124,585
(4)	National index, real estate activity	669	754	874

SOURCE: (1) E. Fisher, *Urban Real Estate Markets*, New York, NBER, 1951, Table A-4; (2) NBER files 0168, Gottlieb, Long Swings study; (3) NBER files 0157, Gottlieb; (4) NBER tabulations, Abramovitz, Long Swings study, series 000661.

[76] See W. H. Person's annual survey, "The Records of Business Since 1875," with his display of the monthly values of eleven basic series (*Forecasting Business Cycles*, New York, Wiley, 1931, p. 123). W. C. Mitchell noted some signs of cyclical slippage—stock exchange sales down, outside clearings off a bit, and industrial consolidations reduced, but "the volume of general business still remained immense" (*Business Cycles*, University of California Press, 1913, p. 64). Note also the low amplitude (3.7) of the 1899-1900 contraction in the fifteen surveyed contractions between 1879-1929 (A. F. Burns and W. C. Mitchell, *Measuring Business Cycles*, New York, NBER, 1946, Table 156, p. 403).

Evaluation of Total Series

Series	Residential Production or Building in 1901-04 Relative to 1896-99
Gottlieb	151.1
Blank (BLS-NBER)	99.2
Long	93.7
Ohio (statewide)	156.2
Riggleman-Isard	143.1

SOURCE: Table 15; NBER series 0147; Colean and Newcomb, *Stabilizing Construction,* p. 231, App. N., Table 2; *Nonfarm Housing Starts,* p. 15.

our derived pattern movement around the decade turning point differs widely from the BLS-NBER one. The difference is largely due to the decision made earlier to step up the rate of decade growth for the 1900's implicit in the BLS-NBER (and the Long) series, and to step down the BLS-NBER decade total for the 1890's. The new pattern of movement around the junction of the two decades follows the pattern of the Ohio and Riggleman-Isard series but diverges sharply from both the Long and BLS-NBER series. No claim can be made that the annual distribution of the altered decade levels is, in any real sense of the word, accurate. The new annual distribution is, however, more consistent with all our information about decade levels, growth trends, and annual patterns of change.

The residential building series developed by the present study can not be accepted as a final solution. The weaknesses of the series and the many improvisations and sheer guesses which entered into their making are self-evident. The calculations were carried through only to test the validity and fruitfulness of the research methods proposed and to sketch out the best possible version of United States nonfarm residential building, given our present knowledge, the limits of this study, and the many attempts at measurement on record.

The next step would be to refine the analysis and test its materials. The following steps—listed not necessarily in the order of their importance —would be important extensions of this work. Critical evaluation of the results reported here will yield additional ones.

1. Since the Ohio building statistics are an important asset in reconstructing the history of residential building in the United States, additional evaluation of them seems desirable. Tests of formal validity, such as checking for internal consistency and plausibility and comparison with independent data, have already been made. It would be helpful to establish by field visits and historical research the character of local records maintained by collecting agents, the responsibility of local data col-

Evaluation of Total Series

lectors, contemporary evaluation of the statistics, and procedures used in administering the statistical reporting system.[77]

2. The Ohio building statistics were compiled for five sample groups of counties which differ in degree of urbanization, size of central cities, and response to economic influences. Tabulation of the Ohio returns—laid out as county totals for some 88 Ohio counties—by still other groupings would permit a more refined system of multipliers by which to develop more accurate residential building estimates. We might also consider grouping Ohio counties into sets, by degree of urbanization and other characteristics to be gleaned from census reports, and then try to project the experience of Ohio county groups to the universe of comparable counties located elsewhere.

3. Apart from the limits imposed by time and money, there is no justification for projecting from Ohio directly to the national aggregate. At the very least, the southern region of the country should be treated separately, because its demography, urbanization, and rates of change preclude its addition to a national projection without allowance for its special characteristics. Complicated southern farm and dwelling statistics need special handling. The census reports of growth of farm enterprises in the South—and hence in the national totals—reflect in part the break-up of the large plantation and the rise of the sharecropper system. Negro housing with its drastically lower level of quality should probably be reported separately from white units or should be stated in terms of "white equivalents."

4. Loss and shrinkage ratios could profitably be studied more thoroughly. The Ohio loss ratios may not be characteristic of eastern seaboard dwelling structures with their greater degree of brick or masonry construction compared with central state dwellings of the period. It is likely that the insurance literature of the nineteenth century with information from other sources would tell the story of losses by fire and disaster. Various government geological agencies collected and published information on the prevalence of stone and brick construction which could be used in developing fire loss ratios. Additional information on estimation of acceptable or plausible loss ratios is badly needed in setting the general level of estimates of housing production.

5. The 1840-60 projection could probably be improved by use of

[77] An effort was made to explore the archives of the state statistical bureau, but all trace of its records had disappeared; county records are, however, available for searching.

Evaluation of Total Series

property-assessment increments available for many states and cities. Real property assessments will reflect, however unevenly, movements of building; and annual tax returns have been compiled in public documents of most states.

6. The work reported in this paper has been carried forward in a parellel study, to be published elsewhere, developing annual estimates from 1850 to 1939 of dollar expenditure upon nonfarm building, both residential and nonresidential.[78] The estimates presented in this study were utilized, together with an independent set of estimates of average per-unit dwelling value in constant dollars, to determine estimates of gross residential new construction. From 1850 to 1912 the per-unit values were derived from the Ohio sources adjusted to exclude farm-dwelling values and to reflect nationwide value levels. The residential values were then projected against a parallel set of values for total nonfarm building. The total values were derived by projection against census bench marks of Ohio building rates to obtain nationwide decade aggregates, which were allocated into yearly returns by use of a set of distributing indexes. For the period 1850 to 1914 the values for nonresidential building were determined residually. The 1915 residual nonresidential values were linked to our established nonresidential building series which had been integrated by the Departments of Commerce and Labor with a detailed network of construction statistics. The fact that linkage could be effected without stretching estimates or adjustments out of shape testifies, it is hoped, to the soundness of the research methods and to the validity of the basic data used here.

---

[78] See my study, "Value of Nonfarm Building," referred to above, n. 62, and available in mimeograph, on request.

Appendix

The Five Adjustments of the 1940 Vintage Report

1. Some 3.2 million converted units were reported in the 1940 Census. Our task was to estimate how many were located in structures built after 1890. If age of structure did not affect conversion, then the proportion of converted units in structures erected after 1890 would be the same as the proportion of total standing stock (87.62 per cent). This represents the maximum conversion allowance. However, the average residential conversion rate on the 1940 standing stock (11.62 per cent) would hardly apply regardless of age. Conversion should be minimal for relatively new structures. Conversion is often thought of as concentrated in the older belt line of central city middle-class residential neighborhoods outmoded by the automobile and the suburban drift. Yet surveys of conversion indicate this is not always true.[79] Detailed surveys for two cities showed that residential age accelerated conversion only mildly in one city, markedly in the other.[80]

Since empirical testimony was inconclusive, an effort was made through multiple regression analysis to see whether age was correlated with rates of conversion, measured by percentage of standing stock converted in 1940. This percentage measured the strength of the tendency to conversion. Detailed experimentation with regression analysis conducted with both state and city returns and use of a wide variety of explanatory variables indicated that conversion was stimulated by age in some areas and for some city-size classes. The final results of the multiple regression anal-

---

[79] *Housing Research,* Apr. 1954, p. 11.

[80] About 65 per cent of the Baltimore conversions, compared with 62.8 per cent of all dwelling units, were made on structures built before 1920. For Norfolk, the corresponding magnitudes were 73 and 42 per cent (*Housing Research,* Nov. 1952, p. 6).

Appendix

TABLE A-1
MULTIPLE REGRESSION PERCENTAGE URBAN DWELLING UNITS CONVERTED, 1939 CENSUS, STATEWIDE BASIS

Variable or Item	Mean Value of Variable Statewide Group 21 States[a]	27 States[b]	Regression Coefficient (Standard Error) 21 States[a]	27 States[b]	Beta Coefficient 21 States[a]	27 States[b]
X_1 Median age structure	29.8	23.9	−.034 (.13)	.305 (.12)	−.04	.51
X_2 Single-family structures as per cent of total	51.9	62.8	−.170 (.09)	.039 (.06)	−.63	.13
X_3 Nonwhite and foreign born population as per cent of total	23.8	24.4	−.475 (.12)	.051 (.06)	−1.27	.14
Constant			34.89	2.55		
R^2			.614	.426		
Percentage dwelling units in structures built in:						
X_4 1910-19	18.6	21.6	.786 (.58)	.103 (.16)	.85	.14
X_5 1900-09	20.1	18.3	.376 (.32)	.478 (.20)	.37	.63
X_6 1890-99	14.0	8.2	.763 (.42)	−.289 (.24)	.60	−.34
X_7 1880-89	7.8	3.4	−.555 (.55)	−.554 (.59)	−.34	−.35
X_8 1860-79	5.7	2.2	.736 (.52)	1.65 (.57)	.71	1.07
Constant			−18.90	2.86		
R^2			.355	.632		

SOURCE: *Sixteenth Census of the United States, 1940, Reports on Housing, II, General Characteristics of Housing by States*, Part 1, Tables 32, 33, 57; *Population*, Vol. I, pp. 19, 30.
[a] Northern states.
[b] Southern and Pacific states.

Appendix

ysis are presented for statewide returns in Table A-1 differentially for 21 northern states and 27 southern and Pacific states. The regression was based on the suppositions that older and single-family buildings would be more prone to conversion on the supply side, and that demand for converted facilities would be generated by a higher proportion of nonwhites and foreign born. Hence, in the first regression model, conversion ratios were regressed against median age of housing stock, per cent of single-family structures in housing stock, and nonwhites and foreign born as a per cent of total population.

This complex of influences "explained" 61 per cent of the variation in conversion rates in northern and central states but only 43 per cent of the variation in southern and Pacific states. Though the "fit" was more successful in the northern states' regression, the explanation was poorer. The signs are all wrong and, in any case, the age fraction showed up as an unimportant influence. The explanatory variables themselves were closely interrelated. The southern and Pacific regressions exhibit a preponderant influence of age on per cent of conversion. Each year of median age was associated with a $.31 \pm .12$ per cent increase in the conversion rate. The relative influence of the other two forces was much smaller.

The influence of age on statewide conversion ratios was isolated in a second regression of conversion percentage on vintage patterns, i.e., the per cent of housing stock originating in periods from 1860 to 1919. Here, again, the southern and Pacific regression showed the preponderant influence, with an "explanation" of 63 per cent of the variance against 36 per cent for the northern group. Northern housing stock built in the 1910's, 1890's, and 1860-79 and southern and western stock erected in the 1900's and in 1860-79 seemed prone to conversion.

While these regressions exhibited a clear influence of age on conversion in the southern and western states, the relationship was more doubtful for the northern states. Hence, it seemed desirable to consider individual northern cities drawn from a relatively homogeneous group of eastern seaboard states. It was decided to permit the influence of age of housing stock to be measured by the per cent of stock found in structures built from 1880 to 1910. A tendency toward conversion of older buildings should be picked up by this measurement. The other factor with a significant relationship was percentage of single-family structures. Experimental regressions indicated the unimportance of minority share of population and average city size. The cities were grouped into those with population between 50,000 and 99,000, and those with more. The results are set forth in Table A-2. Both regressions were equally successful for

Appendix

TABLE A-2

MULTIPLE REGRESSION PERCENTAGE URBAN DWELLING UNITS CONVERTED, 67 EASTERN SEABOARD CITIES OVER 50,000 POPULATION, 1939 CENSUS

Variable or Item	Mean Value Variable, by Size Class Small Cities[a]	Large Cities[b]	Regression Coefficient (Standard Error) 35 Cities	32 Cities	Beta Coefficient 35 Cities	32 Cities
X_1 Single-family structures as per cent of total units	37.1	31.8	.189 (.05)	.058 (.08)	.55	.11
X_2 Units in structures built between 1880 and 1910	44.7	46.7	.10 (.04)	.157 (.04)		.58
Constant			−1.46	−.198		
R^2			.360	.364		

SOURCE: *Sixteenth Census of the United States, 1940, Reports on Housing,* II, *General Characteristics of Housing by States,* Tables 74, 76.

NOTE: Eastern seaboard cities are in Mass., R. I., Conn., N. Y., N. J., Pa, Del., Md.

[a] 35 cities between 50,000 and 99,000 population.
[b] 32 cities over 100,000 population.

Appendix

"fit" (R^2 of .36 and .364), but the primacy varied. Conversion ratios in small cities were preponderantly affected by the relative importance of single-family structures; the ratios in large cities were more affected by the relative share of buildings thirty- to sixty-years old.

Only tentative and guarded conclusions can be drawn from these regression runs. Clearly, there is some influence of age on conversion activity. A 1 per cent increase in the per cent of units in structures 30-60 years old in northern seaboard cities raised the conversion rate by .1 to .16 per cent; and a year of additional median age added .182 (.09) to the nationwide conversion percentage. If age were without influence on conversion activity, then 88.1 per cent of converted properties would be of recorded post-1890 vintage. The assumption here is that 76.7 per cent are actually so recorded or that an additional 362,000 units were converted solely on account of age.

2. "Other dwelling places" was defined in the 1940 Housing Census as inclusive of "tents, trailers, tourist cabins, boats, railroad cars, dugouts, and shacks when occupied by households having no other place of residence." Also included are "garages, warehouses, fruit sheds, barns, caves" when used as places of residence. All these dwelling places, totaling 147,000 units, should be excluded from the standing stock to determine the accuracy of building permit statistics. Transient hotel rooms, missions, cheap overnight lodging houses, dormitories, and institutional accommodations (religious, educational, medical, military, orphanage) are all excluded from the other dwelling places category and from the definition of dwelling units.[84] It seems unlikely that many nonhousekeeping accommodations without plumbing or permanent cooking facilities would have been classified in the 1940 Census as dwelling units, since without such facilities the accommodations would fall under the transient category. In any case, most of the nontransient nonhousekeeping accommodations would either have been converted or would have fallen under the other dwelling places category. It is true that the 1940 Census enumerated 4,712,000 units without running water. That these are bona fide dwelling units with housekeeping facilities is indicated by the fact that 79.5 per cent of these units were one-family structures with little likelihood of converted origin. They were predominantly (90.4 per cent) located out of central cities of over 50,000; some two-thirds of the units were rural nonfarm, widely distributed in the north central and southern states.

[84] See "Introductory Comments," *Sixteenth Census,* 1940, volumes on Housing.

Appendix

Hence it seems fair to infer that few dwelling units enumerated as such in the 1940 Census were of the nonhousekeeping variety, aside from units placed in a "converted" or "other dwelling places" category.

3. The last minus item, farm units transferred to nonfarm stock, is difficult to track down. The numbers of farm units converted to nonfarm stock for the 1940's and 1930's are estimated at 1,250,000 and 91,000, respectively.[85] The rate of conversion increased, partly because of tendencies to urban sprawl in rural surroundings and partly because the number of farm families has declined at a rapid rate since 1940. Since the number of farm families was slowly growing from 1890 onward to 1940, our estimate of 150,000 seems acceptable.

4. The above figure is an order-of-magnitude allowance rounded to the nearest 50,000 mark. The GBW estimates for demolition and other shrinkage between 1890 and 1940 appear unaccountably low, allowing for 1,916,000 units or 9.64 per cent of estimated starts over the same period. This involves an average annual loss rate to stocks of around 0.2 per cent. The vintage report indicates that the 1890 housing stock experienced a compound annual loss rate of 2.09 per cent and an aggregate loss of around 5 million units over the fifty years (see pp. 8, 45). Even if this is cut down by 40 per cent—our maximum allowance for excess estimation—it still exceeds shrinkage allowances implicit in the GBW estimate. However, we may be able to make use of Wickens' estimate for loss by fire and demolition, which for the twenties sets this loss at 43.1 per cent of total losses for the decade, or 25,000 units per year. If we convert this into a decade rate per unit of housing stock and apply this rate to other average decade stocks, we derive an aggregate loss by fire and disaster of 1,005,000 units. This is within 20 per cent of the fire and disaster loss resulting from application of the Wickens' 43 per cent to total loss otherwise estimated by GBW. If we then use vintage aging and mortality patterns for stock of pre- and post-1890 vintage and assume that fire and disaster loss does not respect age, we accumulate a fifty-year total of fire and disaster loss of 693,000 units of stock of post-1890 vintage. If we accept this figure and add a mere token quantity of demolition at a rate of 0.06 per cent per year on the average post-1890 stock, we derive an accumulated total of 300,000 units or an average demolition loss of 6,000 units per year. This yields a total loss of a million units, which does not appear excessive considering the age distribution, rates of urban growth, and changing patterns of land use.

[85] See GBW, p. 373.

Appendix

5. The above is an allocation to post-1890 production of the 7.92 per cent of housing units for which information on year built was not obtained. The 2,352,700 units involved were broadly distributed by regions and structural types. Nonreporting was more characteristic of units located in multifamily structures and possibly for that reason was more characteristic in larger cities and in regions of old settlement. There is a clear progression from 6.0 per cent for single-family detached units to 11.9 per cent for units located in structures containing twenty units or more. As expected, owner-occupied units are nonreported by only 4.5 per cent, tenant-occupied units, 10 per cent. Urban rates as a whole are, however, only slightly higher than nonurban, and there is very little progression through the urban size-class. The principal metropolitan district (PMD) rate (8.5 per cent) compares to urban outside PMD (8.3), all cities over 50,000 (7.2), rural nonfarm (6.0). The regional layout of nonreporting ratios correlates with the age layering and also with emphasis on multi-unit building. Thus the six main regions have conversion rates, nonreporting rates, and median age as follows:

States	Conversion Rate, Urban	Nonreporting Rate, All Units	Median Age, Urban
	(per cent)		
New England (all units)	9.3	13.76	34.5
Middle Atlantic	10.6	12.70	29.2
East north central	13.1	5.42	27.6
West north central	15.2	4.67	29.8
South	14.2	5.74	21.6
West	10.7	4.40	19.5

SOURCE: 16th Census, *Housing,* II, 1943, Tables 33 and 57.

Different forces are at work in the regions, so that no clear association prevails between conversion, nonreporting, and median age. The regional patterns indicate that older buildings in New England and in the Middle Atlantic states were more likely to be unreported for vintage. But, in the four other regions, age and nonreporting did not correlate positively. We tested the aggregate correlation indicated in the Northeast by inquiry into patterns of behavior of individual cities. Scatter diagrams relate reporting percentage in eastern cities to median age and, separately, to percentage of units built between 1880 and 1910. A wide scatter was indicated with only a slight tendency for an inverse relationship of higher age with reporting percentage. It is doubtful that multiple correlation would sustain this result and, in any case, considerable experimentation would be needed to design a useful model. At least in two regions with a

Appendix

wide age spectrum, these diagrams indicate that advanced age in measurable form did not correlate positively with nonreporting.

Notwithstanding this negative result in the two regions, it would be extreme to assume that 87.6 per cent of nonreporting units were of post-1890 vintage. The most recent decades can surely be recalled more easily, the more remote with difficulty. We allow for this tendency generously—in view of the evidence—by crediting the same tendency for association of age with nonreporting that we allowed for conversion. This credits 1,805,000 nonreporting units or 76.7 per cent to post-1890 vintage.